CUNARD PORTRAITS

144 scale line drawings of ships of the Cunard fleet

BY JOHN H. ISHERWOOD

Research and design by
Kevin O'Donoghue and Michael Crowdy

INTRODUCTION

THIS BOOK is published by the World Ship Society to commemorate the 150th anniversary of "Cunard", and as a tribute to the memory of Captain John H. Isherwood.

It includes virtually all the drawings which he made of the ships owned by The Cunard Steam-Ship Company p.l.c. and its predecessors and they form a striking pictorial record of the development of passenger shipping during the past 150 years.

The drawings, which were made with meticulous care during a period of over thirty years from the early 1950s to the mid-1980s, are all to the uniform scale of 1:1200 (one inch to one hundred feet) and form part of the John H. Isherwood Collection of over 6,000 of his drawings of ships which is to be maintained in Southampton and will be accessible to people who wish to refer to them.

It will be noted that in a number of cases the funnels of the ships are not coloured red. Except in the case of CUNARD ADVENTURER, which had a white funnel, this is because the colouring had been omitted from the original drawing which could have been intended for publication.

A number of the ships in the book were the subject of articles by Captain Isherwood in his series "Steamers of the Past" in the magazine "Sea Breezes" and for the benefit of readers who may wish to refer to these detailed histories, the ships concerned are marked with an asterisk in the index.

The synoptic history of Cunard, and the data and brief histories given below the pictures of each ship, are based upon information in the Central Record of the World Ship Society and in particular the records of the late Noel Bonsor, author of "North Atlantic Seaway", who bequeathed his collection to the Society. A brief note about the World Ship Society appears on page 64.

In conclusion, we would like to place on record our thanks to Mrs. Margaret Isherwood and Mrs. Joan Sims, whose kindness and helpfulness has made this book possible.

KEVIN O'DONOGHUE
MICHAEL CROWDY

KENDAL, JUNE 1990

Published in 1990 by the World Ship Society
28 Natland Road, Kendal, Cumbria LA9 7LT, England

ISBN 0 905617 57 6

Front cover: BRITANNIA (1840), SERVIA (1881), MAURETANIA (1906), QUEEN ELIZABETH 2 (1969)
Rear cover: QUEEN MARY (1936)

A VERY BRIEF HISTORY OF CUNARD

ONE HUNDRED and fifty years ago, on the 4th July 1840 the first "Cunarder", the wooden-hulled paddle steamer BRITANNIA, sailed from Liverpool on her maiden voyage to Halifax, Nova Scotia (Canada), and Boston in the United States, carrying mails and sixty-three passengers.

At that time the steamship was in its early infancy, but successful voyages from England to New York by the British paddle steamers GREAT WESTERN and SIRIUS during 1838 had decided the British Admiralty to invite tenders during November that year for the carriage by steamship of the mail between England and North America.

Samuel Cunard, an established businessman and shipowner in Halifax, N.S., travelled to London early in 1839 and put forward a tender based on a fortnightly service between Liverpool and Halifax, to be maintained by three ships, and with a "feeder" service to Boston.

The tender was accepted, but in the light of subsequent advice from Robert Napier, the Glasgow shipbuilder whom Cunard had contracted to build the ships, it was decided that four, larger, ships would be required to maintain the service.

This entailed greater expense than Cunard had envisaged and so, in association with the Glasgow shipowners George Burns and Charles and David MacIver, and others, he formed the British and North American Royal Mail Steam Packet Company. The Company had a capital of £270,000 of which Cunard subscribed £55,000.

Further negotiations were conducted with the Admiralty and early in 1840 the contract finally agreed, for a period of seven years at an annual payment of £56,000, was for a service to and from Boston, calling intermediately at Halifax, to be maintained by four ships giving a fortnightly sailing, except in the months of November to March, when it was to be monthly.

By the Spring of 1841 it was apparent that costs were considerably exceeding income and later in the year the mail subsidy was increased to £81,000 per year, on condition that a fifth ship was built. This was HIBERNIA, a faster vessel, and the first Cunarder to sail eastbound from New York, on 1st January 1848. Her sister CAMBRIA, built to replace COLUMBIA which had been lost in July 1843, sailed from Liverpool on the same day bound via Halifax for New York. The frequency of the service was now doubled and the United States terminal alternated between Boston and New York.

Two years later severe competition arrived on the route with the Collins Line's four fast ships but in the same year Cunard were awarded a subsidiary mail contract from New York to the British West Indies, although this service was not, in fact, introduced until late 1858 following the Anglo-French war against Russia.

In 1852 Cunard established a service from Liverpool to Mediterranean ports, the initial sailing being taken on 17th July by BRITISH QUEEN, and towards the end of the 1850's the British & Foreign Steam Navigation Company was formed by Cunard, Burns and MacIver to own and operate the ships concerned. There was, however, a measure of inter-change, as it was one of the Mediterranean ships, KARNAK, that inaugurated the West Indies service.

The same year saw the entry into service of the first iron-hulled screw-propelled Cunarders, ANDES and ALPS, whilst the last of the Company's wooden-hulled paddle steamers, ARABIA, sailed on her maiden voyage on the 1st January 1853. PERSIA, completed in 1856, had an iron hull, but retained paddle propulsion and regained the speed record from the Collins Line, subsequently relinquishing it to the 1862-built SCOTIA, the last of the Cunard paddle steamers.

In 1860 ETNA and JURA were the first Cunard ships to be fitted for the conveyance of steerage passengers, whilst 1862 saw the delivery of CHINA, the first screw-propelled ship to be built for the Cunard mail service. The last mail sailing to Halifax took place in December 1867, the contract then being taken over by the Inman Line.

Two ships destined to have long lives were delivered in 1870: BATAVIA, the first Cunard ship to have compound steam engines, which lasted until 1924 when she was wrecked while sailing under the Japanese flag, and PARTHIA, which remained in service as a cargo steamer until 1952 and then served for another four years as a barge before being broken up in Japan.

Following the advent of the White Star Line in 1871, the Cunard fleet was somewhat over-shadowed and in order to raise the capital necessary to create a competitive fleet, the Cunard Steam Ship Company Limited was formed in 1878 with a capital of £2,000,000 of which £1,200,000 was issued to the Cunard, Burns and MacIver families in respect of their interests in the British & North American Royal Mail Steam Packet Company and the British & Foreign Steam Navigation Company. £800,000 was subscribed by the public.

The mail subsidy had expired at the end of 1876 and henceforth payment was on the basis of weight of mail carried.

Apart from the second-hand OREGON, the next notable Cunarders were UMBRIA and ETRURIA, which held the North Atlantic speed record for nearly four years up to 1889 and were the last of the single-screw record breakers. Two years later, in 1891, the twin-screw triple-expansion engined CAMPANIA and LUCANIA were ordered for the New York service and when delivered in 1893 they quickly regained the speed record from the Inman liners (CITY OF PARIS and CITY OF NEW YORK) which had taken it from ETRURIA. They were to hold it until the German KAISER WILHELM DER GROSSE entered service at the end of 1897.

For the Liverpool-Boston service the cargo and cattle carriers SYLVANIA and CARINTHIA were delivered in 1895 and were followed in 1898 by the larger ULTONIA. All had limited passenger accommodation but in 1898 ULTONIA was refitted to carry 675 steerage class passengers. CARINTHIA was lost in 1900 whilst on Government service, but in the same year the quadruple-expansion engined passenger ships IVERNIA and SAXONIA came into service and were noteworthy for having the tallest funnels to be fitted in a ship. On her April 1901 sailing from Liverpool the latter carried the then record total of 2,260 fare-paying passengers.

In 1903 they were joined on the route by the slightly smaller CARPATHIA, which had accommodation for a total of 1,704 passengers, but after just one voyage she was transferred to the Liverpool-New York service, and in 1904 to the Trieste-New York service which had developed following Cunard's appointment as official agents for Hungarian emigration. Her accommodation was then increased to cater for a total of 2,550 passengers, 2,250 of them in steerage. SLAVONIA and PANNONIA were purchased for this service and when ULTONIA was also transferred to it, her passenger accommodation was increased again, to a total of 2,220, of which 2,100 were steerage berths.

In 1902 the American-owned International Mercantile Marine Company acquired the main British rival to Cunard, the White Star Line, as well as a number of other companies trading on the North Atlantic, and the British Government undertook to loan Cunard the cost of building two large, fast passenger liners on condition the company remained in British control. The liners, which were ordered in 1905, and sailed on their maiden voyages in 1907, were LUSITANIA and MAURETANIA. The former was sunk during the First World War, but the latter became one of the best-loved ships of all time, holding the "Blue Riband" — as the speed record for the North Atlantic had come to be known — from 1909 until 1929.

Until 1905, screw-propelled passenger liners on the North Atlantic had been powered by reciprocating engines, but these two express liners were fitted with steam turbines, which had been successfully tested by Cunard in the "intermediate" liner CARMANIA which entered service at the end of 1905, less than nine months after the very first North Atlantic turbine-powered liner, Allan Line's VICTORIAN.

MAURETANIA's fastest west-bound passage was made at a speed of 26.06 knots, and east-bound (as late as 1924) at 26.25 knots.

She was followed in 1914 by the larger, but somewhat slower, AQUITANIA which was intended to maintain a three-ship express service between Liverpool and New York in company with her and LUSITANIA.

FRANCONIA and LACONIA had meantime been built for the Boston service, and ANDANIA and ALAUNIA for the Canadian trade which Cunard had re-entered in 1911 when they acquired the Thomson Line passenger ships from Cairn Line who had commenced a London-Southampton-Quebec-Montreal service the previous year.

This acquisition was, in fact, the first of a number effected by Cunard in the period up to 1921: The Anchor Line, of Glasgow, who traded to New York from Glasgow and the Mediterranean were taken over in 1912 and were followed in 1916 by both Canadian Northern Steamships Ltd whose ships had sailed pre-war between Avonmouth and Canada, and the Commonwealth and Dominion Line, who were important operators in the Australasian trade. In 1916 also, Anchor Line joined with Donaldson Brothers Ltd. to form Anchor-Donaldson Ltd., to which the goodwill and four ships of the Donaldson Line's Canadian passenger service were transferred. This, in effect, gave Cunard entry to the Glasgow-Canada trade. The last of the acquisitions was in 1919/21 when Cunard took over from Anchor Line control of Thos. & Jno. Brocklebank Ltd., who were long-established in the Indian trade.

The Company suffered substantial losses during the war, but were able to resume a regular Liverpool-New York service before the end of 1918. Cunard having announced that they were going to replace the German Lines on the Southampton — New York route, AQUITANIA made the first sailing in June 1919 from Southampton, which thereafter became the English terminal for the express service to New York, the service being maintained from 1921 by her, MAURETANIA and the ex- German BERENGARIA.

In addition to the transatlantic cargo services, the other principal Cunard services in the 1920's and 30's were from Liverpool to Boston and/or New York, Liverpool to Quebec and Montreal and London/Southampton to Quebec and Montreal. FRANCONIA and CARINTHIA, built for the Liverpool/New York service, were specifically designed so that they could undertake cruising in the winter season and in her later years MAURETANIA, among others, was similarly employed.

In 1929 the new German liner BREMEN captured the Blue Riband and in May the following year Cunard announced that they had placed an order for the ship that eventually became the QUEEN MARY. Laid down on the 27th December 1930, work on her was suspended twelve months later owing to the Depression. In December 1933 the Government announced that it was prepared to advance £3 million to Cunard to enable the ship to be completed and £1.5 million as working capital. A further £5 million was to be made available later if a sister ship was built. This financial deal was made conditional upon Cunard merging with the other major British company on the North Atlantic, the White Star Line. The resulting company, Cunard White Star Ltd., was registered on 10th May 1934, with a capital of £10 million, of which Cunard Steam Ship Company Ltd. held 62%.

QUEEN MARY was launched five months later and within three months of her maiden voyage in May 1936 had recaptured the North Atlantic record with crossings in each direction at speeds in excess of 30 knots. This was the first time that 30 knots had been exceeded and her record stood until the advent in 1952 of the 35-knot UNITED STATES.

The keel of her sister ship was laid in December 1936 and was launched as QUEEN ELIZABETH in September 1938. Owing to the war, she was destined not to enter commercial service until October 1946, although during the war she and QUEEN MARY had served as troopships, each carrying up to 15,000 troops at a time.

The second MAURETANIA, which had the distinction of being the largest merchant ship yet built in England, sailed on her maiden voyage from Liverpool to New York in June 1939 and later she, too, served as a troopship. She survived the war, but several ships of the pre-war fleet were lost and others were sold to the Admiralty.

After the war, the first newly-built Cunarder was the cargo ship ASIA, which sailed on her maiden voyage in April 1947, being followed in August by MEDIA, which was the first post-war built passenger liner on the North Atlantic.

During 1947, the Cunard Steam Ship Co. Ltd. acquired, for £2 per £1 share, the Government's 38% shareholding in Cunard White Star Ltd. and in 1949 amended their own name to The Cunard Steam-Ship Co. Ltd.

January 1949 saw the maiden voyage of the "dollar earner" CARONIA, which had been designed for summer service on the North Atlantic and cruising the rest of the year but when SAXONIA and her three sisters were delivered between 1954 and 1957 for the Liverpool-Canada services they were to be the last Cunard ships built for year-round passenger liner service. By the end of the decade plans for a new 75,000 ton express liner were under discussion, but already competition from the air was making itself felt in the business travel market.

This competition caused the plans for the liner to be abandoned but in 1964 it was decided to build the ship, suitable for transatlantic service in summer and for cruising in winter, that was completed five years later as the pride of the Cunard fleet, QUEEN ELIZABETH 2.

Events during the 1960s decided Cunard that so far as passenger ships were concerned, the future lay in a cruise-orientated operation. Accordingly, the Liverpool-New York direct service ended in 1966 and the St. Lawrence services in 1967, and when SYLVANIA sailed from Liverpool on the 30th November 1967 for Halifax and New York, it was to be the last Cunard passenger sailing on the route which they had established 127 years before. The same month CARONIA was withdrawn from service, while the month previously QUEEN MARY had sailed for Long Beach, California, to become a maritime museum and hotel ship. When QUEEN ELIZABETH was withdrawn from service in November 1968 only CARMANIA and FRANCONIA remained and they lasted, in a cruising role, until 1971.

In contrast with this run-down of the passenger side, at least until QUEEN ELIZABETH 2 was delivered in 1969, there was expansion and diversification of the cargo operations. In 1964 the tanker owners H. E. Moss & Co. were acquired and when Associated Container Transportation (Australia) Ltd. was formed Cunard, through Port Line, had a 42.5% interest, which they subsequently increased. In 1967, when Cunard-Brocklebank Ltd. was formed to manage the cargo vessels in the fleet, Cunard also took a 20% interest (sold in 1989) in Atlantic Container Line Ltd., and had two ships built for ACL service. Later, in April 1968, they acquired Offshore Marine Ltd., a company engaged in servicing the offshore oil industry in the North Sea.

Following the delivery of the new ACL ships, the Cunard break-bulk cargo services to North America ceased in 1970, the U.K.-Mediterranean cargo service also being terminated in the same year.

On the 24th August 1971 The Cunard Steam-Ship Co. Ltd. was purchased by Trafalgar House Investments Ltd. for £26,000,000, thus ending their independent existence. They remain part of the Trafalgar House Group.

In the latter part of 1971, the first of two new Dutch-built cruise ships was delivered but, like the eight bulk carriers built shortly afterwards in Spain, they did not remain long in the fleet.. Much more successful, however, were the two Danish-built cruise ships, CUNARD COUNTESS and CUNARD PRINCESS, which were delivered in 1976-77 and continue in Cunard service.

1976 also saw the acquisition of ten refrigerated cargo carriers which remained in service until 1985-86; Offshore Marine Ltd. and its fleet was sold to Zapata at the beginning of the new decade.

The next major development was the acquisition in 1984 of the Norwegian cruise ships SAGAFJORD and VISTAFJORD, which complete the pictorial history in this book. They were joined in the cruise fleet in 1986 by the chartered Norwegian SEA GODDESS I and SEA GODDESS II, giving Cunard today a total of seven passenger vessels.

In 1987 Trafalgar House acquired Ellerman Lines p.l.c. and the managers of the cargo division became Cunard Ellerman Shipping Services Ltd, Cunard thereby re-entering the Mediterranean trade.

Today, in 1990, Cunard is part of a group with wide interests in the hotel and leisure markets, it owns one of the leading cruise ship fleets in the world, a 58,000 ton container carrier, tankers and, with Ellerman, has a substantial cargo trade to and from the Mediterranean.

BRITANNIA (1840-1849) Wooden Paddle Steamer.
1135 gross, 619 net. 207 x 34 feet. Passengers: 115 1st class.
2-Cylinder Side Lever steam engine manufactured by Robert Napier and Sons, Glasgow. Speed 9 knots.

Launched 5.2.1840 by Robert Duncan and Co., Greenock, for the company; the first vessel owned. Sailed 4.7.1840 on her maiden voyage from Liverpool to Halifax and Boston. Sold 1849 to the Navy of the German Confederation for service as a frigate, she was renamed BARBAROSSA, retaining this name when transferred in 1852 to the Prussian Navy. In 1880 she was sunk as a target ship.
Sisterships: ACADIA, CALEDONIA, COLUMBIA.

HIBERNIA (1842-1850) Wooden Paddle Steamer.
1422 gross, 791 net. 207 x 35 feet. Passengers: 120 1st class.
2-Cylinder Side Lever steam engine manufactured by Robert Napier and Sons, Glasgow. Speed 9 knots.

Launched 8.9.1842 by Robert Steele and Co., Greenock, for the company. Sold 1850 to the Spanish Navy, converted to a frigate, and renamed HABANOIS. Lost in 1868.
Sistership: CAMBRIA.

EUROPA (1847-1867) Wooden Paddle Steamer.
1834 gross, 992 net. 251 x 38 feet. Passengers: 140 1st class.
2-Cylinder Side Lever steam engine manufactured by Robert Napier and Sons, Glasgow. Speed 10 knots.

Launched 22.9.1847 by John Wood and Co., Port Glasgow, for the company's North Atlantic services. On 27.6.1849 she collided in fog in the North Atlantic with the American emigrant brig CHARLES BARTLETT which sank with the loss of 135 lives. The EUROPA sustained little damage, and having picked up survivors proceeded with her voyage. In 1867 she was sold and reduced to a sailing ship.
Sisterships: CANADA, AMERICA and NIAGARA.

EUROPA, after removal of her mizzen mast in 1853

ASIA (1850-1867) Wooden Paddle Steamer.
2226 gross, 1214 net. 266 x 40 x 27 feet. Passengers: 130 1st class, 30 2nd class.
2-Cylinder Side Lever steam engine manufactured by Robert Napier and Sons, Glasgow. Speed 12 knots.

Launched 31.1.1850 by Robert Steele and Co., Greenock, for the company's North Atlantic services. In 1867 she was sold to Port Glasgow owners and her engine was removed. She was destroyed by fire 2.12.1876 at Bombay whilst loading cotton.
Sistership: AFRICA.

ARABIA (1852-1864) Wooden Paddle Steamer.
2402 gross, 1474 net. 284 x 41 x 27 feet. Passengers: 180 1st class.
2-Cylinder Side Lever steam engine manufactured by Robert Napier and Sons, Glasgow. Speed 12 knots.

Launched 21.6.1852 by Robert Steele and Co., Greenock, for the company's North Atlantic services. She had been laid down as PERSIA, but was renamed prior to her launch. This was the first Cunard liner to have two funnels, and was reported to have had an excessive coal consumption — some 120 tons per day. In 1854 she was taken up for service as a transport during the Crimean War. In 1864 she was sold, and her engines were removed, thus reducing her to a sailing ship.
Sistership: ARABIA (sold whilst under construction to Royal Mail Steam Packet Company).

ALPS (1852-1859) Iron Steamship.
1440 gross, 773 net. 236 x 33 x 24 feet. Passengers: 62 1st class, 122 2nd class.
2-Cylinder Geared Beam steam engine manufactured by the shipbuilder. Speed 9 knots.

Launched 28.9.1852 by Wm. Denny and Brothers, Dumbarton, for the company's North Atlantic services. This class were the first iron screw steamers constructed for the company. In 1854 she was taken up for service as a transport during the Crimean War, and in 1859 was sold to the Spanish Government and renamed MANDINGO.
Sistership: ANDES.

BRITISH QUEEN (1851-1898) Iron Steamship.
773 gross, 565 net. 187 x 28 x 17 feet. Passengers: 71 1st class, 166 3rd class.
2-Cylinder Direct Acting Vertical steam engine manufactured by Caird and Co., Greenock. Speed 9 knots.

Launched 8.6.1849 by Wm. Denny and Brothers, Dumbarton, for Brownlow, Pearson and Co., Hull. Purchased 1851 for the company's Mediterranean service, but not renamed. In 1854 she was taken up for service as a transport during the Crimean War. Her original engine was removed in 1878 and replaced by a compound steam engine manufactured by J. Jack and Co., Liverpool. In 1898 she was sold and in 1899 she was broken up at Preston by Thos. Ward.

TAURUS (1853-1859) Iron Steamship.
1127 gross, 812 net. 210 x 29 x 24 feet.
2-Cylinder Geared Beam steam engine manufactured by Tulloch and Denny, Dumbarton. Speed 9 knots.

Launched 21.2.1853 by Wm. Denny and Brothers, Dumbarton, for the company's Mediterranean service. In 1859 she was sold to the Spanish Government. Sisterships: TENERIFFE and KARNAK.

PERSIA (1855-1868) Iron Paddle Steamer.
3300 gross. 376 x 45 x 31 feet. Passengers: 200 1st class, 50 2nd class.
2-Cylinder Side Lever steam engine manufactured by the shipbuilder. Speed 13½ knots.

Launched 3.7.1855 by Robert Napier and Sons, Glasgow, for the company's North Atlantic services. When launched she was described as the world's largest ship, and drew a crowd of over 55,000 people. In 1867 she was laid up and her engine was removed. Sold to Glasgow owners, it was their intention to return her to service as a sailing ship but the venture collapsed. In 1872 she was broken up in the River Thames.

DAMASCUS (1856-1861) Iron Steamship.
1214 gross, 825 net. 253 x 31 x 22 feet. Passengers: 40 1st class, 300 3rd class.
2-Cylinder Inverted Direct Acting steam engine manufactured by the shipbuilder. Speed 10 knots.

Launched 30.9.1856 by Wm. Denny and Brothers, Dumbarton, for the company's Mediterranean service. She had been laid down as ST. GEORGE for Captain D. R. MacGregor of Leith, but was cancelled at an early stage. In 10.1861 she was sold back to Peter Denny in part payment for the newbuilding MOROCCO. He resold her in 1.1862 to Allan Line of Glasgow, and in 1865 she was renamed CORINTHIAN. In 1870 she was extensively rebuilt, lengthened, and her engines were compounded by Greenock Foundry Co., Greenock. In 1881 she was sold to Italian owners as GENOVA, and was subsequently renamed G. LANZA and GIUSEPPE GARIBALDI before passing to Turkey as FOULAZI OSMANI. She was eventually broken up in 1912 as SAKARIAH.

PALESTINE (1858-1872) Iron Steamship.
1800 gross. 276 x 36 x 24 feet. Passengers: 32 1st class, 280 steerage.
2-Cylinder Geared Oscillating steam engine manufactured by Robert Napier and Sons, Glasgow. Speed 10 knots.

Launched 1858 by Robert Steele and Co., Greenock, for the British and Foreign Steam Navigation Company; a company which had been formed by Cunard and his partners to operate a Mediterranean service. In 1860 she was transferred to the Liverpool-New York service and thus to the British and North American Royal Mail Steam Packet Company. In 1872 she was sold to William Henry Jones of Liverpool, lengthened and the engine compounded. Sold in 1881 to E. S. Jones of Brighton, she remained on the North Atlantic until broken up in 1896. Sisterships: OLYMPUS, MARATHON, ATLAS, HECLA, KEDAR, SIDON and MOROCCO.

MARATHON (1860-1898) Iron Steamship.
1784 gross. 274 x 36 x 25 feet. Passengers: 70 1st class, 850 3rd class.
2-Cylinder Geared Oscillating steam engine manufactured by the shipbuilder. Speed 10 knots.

Launched 1860 by Robert Napier and Sons, Glasgow, for the company's Mediterranean service. In 1873 she underwent considerable modernisation, including lengthening, and her engines were compounded by J. and G. Thomson, Glasgow. Sold to Workman, Clark and Co., Belfast, in 1898, she was broken up in Italy as CYPRIA shortly afterwards.
Sisterships: PALESTINE, OLYMPUS, ATLAS, HECLA, KEDAR, SIDON and MOROCCO.

OLYMPUS (1860-1881) Iron Steamship — as lengthened.
1794 gross. 276, later 340 x 36 x 25 feet. Passengers: 70 1st class, 900 3rd class.
2-Cylinder Geared Oscillating steam engine manufactured by the shipbuilder. Speed 10 knots.

Launched 10.1.1860 by J. and G. Thomson, Glasgow, for the company's Mediterranean service. Lengthened and extensively modernised in 1872, she continued to serve on various services until resold in 1881 to her builders. It was common practice at that time for owners to offer vessels in part payment for new buildings, and in this instance she was used to offset part of the cost of PAVONIA. Sold again in 1882 to A. Galbraith, Glasgow, she was again resold to J. and G. Thomson in 1889 before being finally sold for demolition in 1891.
Sisterships: PALESTINE, MARATHON, ATLAS, HECLA, KEDAR, SIDON and MOROCCO.

CALABRIA (1869-1876) Iron Steamship.
2902 gross, 1730 net. 331 x 42 x 29 feet. Passengers: 200 1st class, 60 2nd class.
2-Cylinder Single Expansion steam engine manufactured by the shipbuilder. Speed 12 knots.

Launched 10.6.1857 by J. and G. Thomson, Glasgow, as AUSTRALASIAN for European and Australian Royal Mail Company. She had in fact been laid down as AUSTRALIAN, but the name was changed early during construction in order to avoid confusion with the vessels of the Australian Navigation Company. Following the financial collapse of her owners, she was purchased in 1859 by British and North American Royal Mail Steam Packet Company, but not renamed. In 1869/70 she was extensively modernised, re-engined and the fore funnel was removed. It was at this time that she was renamed CALABRIA. Sold 1876 to Telegraph Construction and Maintenance Co., London, she was converted to a cableship and was finally broken up in 1897 in Holland.

KEDAR (1860-1897) Iron Steamship.
1876 gross, 1215 net. 275 x 36 x 25 feet. Passengers: 30 1st class, 494 3rd class.
2-Cylinder Geared Oscillating steam engine manufactured by the shipbuilder.
Speed 10 knots.

Launched 17.8.1860 by Wm. Denny and Brothers, Dumbarton, for the company's Mediterranean service. Her engine was compounded in 1873 by J. Jack, Rollo and Co. at Liverpool. She was sold 1897 to Italian shipbreakers at Genoa.
Sisterships: OLYMPUS, PALESTINE, MARATHON, ATLAS, HECLA, SIDON and MOROCCO.

SIDON (1861-1885) Iron Steamship.
1872 gross, 1199 net. 275 x 36 x 25 feet. Passengers: 69 1st class, 550 3rd class.
2-Cylinder Geared Oscillating steam engine manufactured by the shipbuilder.
Speed 10 knots.

Launched 20.8.1861 by Wm. Denny and Brothers, Dumbarton, for the company's Mediterranean service. Her engine was compounded in 1873 by J. Jack, Rollo and Co. at Liverpool. Her career came to an end when she was wrecked on 27.10.1885 near Malpica in Spain. She had been on a voyage from Liverpool to Constantinople with a general cargo.
Sisterships: OLYMPUS, PALESTINE, MARATHON, ATLAS, HECLA, KEDAR and MOROCCO.

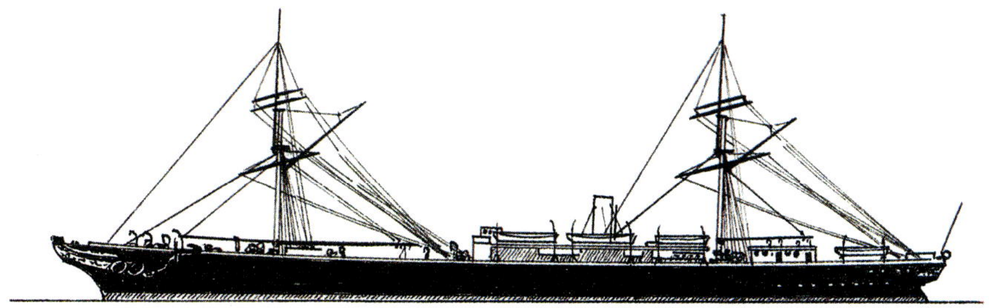

SCOTIA (1861-1879) Iron Paddle Steamer.
3871 gross, 2125 net. 379 x 47 x 30 feet. Passengers: 573 1st class.
2-Cylinder Side Lever steam engine manufactured by the shipbuilder. Speed 14 knots.

Launched 25.6.1861 by Robert Napier and Sons, Glasgow, for the company's North Atlantic services. In 1879 she was sold to the Telegraph Construction and Maintenance Co. and sent to Laird Brothers at Birkenhead for conversion to a cableship. She was extensively altered, re-engined and re-rigged. On 11.3.1904 she ran aground on Catalan Bank, near Guam, broke in two and became a total loss.

CHINA (1861-1880) Iron Steamship.
2638 gross, 1614 net. 326 x 40 x 27 feet. Passengers: 268 1st class, 771 2nd class.
2-Cylinder Geared Oscillating steam engine manufactured by the shipbuilder. Speed 12 knots.

Launched 8.10.1861 by Robert Napier and Sons, Glasgow, for the company's North Atlantic services. Her engine was compounded in 1874 by Barclay, Curle and Co., Glasgow, and in 1879 she was taken up as a transport in the Zulu War. Sold in 1880 to Spanish owners, she was renamed MAGALLANES, and was then sold to Norway as THEODOR. Her new owners removed her engine and had her converted to a four-masted barque. On 2.3.1906 she sailed from Tampa for Yokohama and went missing with all hands.

ALEPPO (1864-1909) Iron Steamship.
2057 gross, 1399 net. 292 x 38 x 26 feet. Passengers: 46 1st class, 593 3rd class.
2-Cylinder Geared Oscillating steam engine manufactured by the shipbuilder. Speed 11 knots.

Launched 1.11.1864 by J. and G. Thomson, Glasgow, for the company's Mediterranean service. In 1879/80 she was extensively rebuilt and fitted with a new engine by J. Jack and Co. at Liverpool. This was again replaced in 1890 when a triple expansion steam engine manufactured by J. Howden and Co., Glasgow, was fitted. After a long career in the service of the company she was sold 1909 to Thos. Ward for demolition at Preston.
Sisterships: TARIFA and MALTA.

JAVA (1865-1878) Iron Steamship.
2696 gross, 1761 net. 337 x 42 x 27 feet. Passengers: 300 1st class, 800 3rd class.
2-Cylinder Inverted Direct Acting steam engine manufactured by the shipbuilder. Speed 12 knots.

Launched 24.6.1865 by J. and G. Thomson, Glasgow, for the company's North Atlantic services. The engine was compounded in 1877 by Fawcett, Preston and Co., Liverpool, and in 1878 she was sold to Soc. Anon. de Navigation Belge-Americaine (Red Star Line), Antwerp, and was renamed ZEELAND. Her engine was removed following sale to French owners in 1889, whereupon she was renamed ELECTRIQUE. Returning to British owners in 1892, she became John Herron and Co.'s LORD SPENCER. On 9.4.1895 she sailed from San Francisco for Queenstown with a cargo of grain in sacks and went missing with all hands.
Sistership: CUBA.

RUSSIA (1867-1880) Iron Steamship.
2960 gross, 1710 net. 358 x 43 x 27 feet. Passengers: 235 1st class.
2-Cylinder Inverted Direct Acting steam engine manufactured by the shipbuilder. Speed 14 knots.

Launched 20.3.1867 by J. and G. Thomson, Glasgow, for the company's North Atlantic services. In 1878 she was taken up as a transport in the Zulu War, and in 1880 was sold to the Red Star Line, Antwerp. She was then extensively modernised by Harland and Wolff at Belfast and her engine was compounded. Returning to service later that year, she was renamed WAESLAND. In 1889 a new triple expansion engine was fitted at Glasgow. Her end came on 5.3.1902 when she was lost in collision with the Houston liner HARMONIDES off Anglesey.

SAMARIA (1868-1902) Iron Steamship.
2605 gross, 1695 net. 320 x 39 x 26 feet. Passengers: 130 1st class, 800 3rd class.
2-Cylinder Inverted Direct Acting steam engine manufactured by the shipbuilder. Speed 12 knots.

Launched 4.7.1868 by J. and G. Thomson, Glasgow, for the company's North Atlantic services. Her engine was compounded in 1878 by J. Jack and Co., Liverpool and she was sold in 1902 to Italian shipbreakers at Genoa.
Sistership: SIBERIA.

ALGERIA (1870-1881) Iron Steamship.
3428 gross, 2193 net. 363 x 42 x 26 feet. Passengers: 200 1st class, 1054 3rd class.
2-Cylinder Compound Inverted steam engine manufactured by the shipbuilder. Speed 13 knots.

Launched 12.7.1870 by J. and G. Thomson, Glasgow for the company's North Atlantic services. In 1881 she was sold to the Red Star Line, Antwerp, and renamed PENNLAND. Before return to service her engine was compounded by J. Jack and Co. at Liverpool. In 1903 she was sold to Italian shipbreakers.
Sistership: ABYSSINIA.

BATAVIA (1870-1884) Iron Steamship.
2553 gross, 1628 net. 327 x 39 x 26 feet. Passengers: 150 1st class, 800 3rd class.
2-Cylinder Compound Inverted steam engine manufactured by the shipbuilder. Speed 12 knots.

Launched 1.2.1870 by Wm. Denny and Brothers, Dumbarton, for the company's North Atlantic services. One report suggests that Cunard purchased her on the stocks, but this has not been confirmed. During 1880 sailed to Bombay on an experimental voyage for the company and stranded at Perim. Refloated with some difficulty, she was returned to the North Atlantic service after repair, and the Indian service was discontinued. In 1884 she was sold together with PARTHIA to Sir William Pearce, who controlled the Glasgow shipbuilder John Elder and Co. Cunard had the UMBRIA and ETRURIA under construction there, and the older steamers were offered as part payment. After fitting her with a four cylinder compound steam engine she was returned to service and sold 1888 to Guion S.S. Co. Ltd., London. In 12.1892 she was renamed TACOMA and was sold 1898 to the North American Mail S.S. Co. of Tacoma. She passed through various American owners until captured by the Japanese on 4.2.1905 when on a voyage from Tacoma to Vladivostock with food supplies. Condemned as a prize, she was renamed SHIKOTAN MARU and ended her days owned by Miyata Momosuke of Kobe. She ran ashore 3.10.1924 off Shaweishan when on a voyage from Tsingtau to Shanghai with a cargo of coal. Although refloated 7.11.1924, she was declared a constructive total loss and was broken up.

As built

As VICTORIA

PARTHIA (1870-1884) Iron Steamship.
3167 gross, 2035 net. 360 x 40 x 34 feet. Passengers: 150 1st class, 1031 3rd class.
2-Cylinder Compound steam engine manufactured by Denny and Co., Dumbarton. Speed 12 knots.

Launched 10.9.1870 by Wm. Denny and Brothers, Dumbarton, for the company's North Atlantic services. In 1884 she was sold together with BATAVIA to Sir William Pearce, who controlled the Glasgow shipbuilder John Elder and Co. Cunard had the UMBRIA and ETRURIA under construction there, and the older steamers were offered as part payment. After Elders had fitted her with a triple expansion steam engine she was returned to service, but in 1891 it was decided to refit her extensively, and her appearance was much changed. After refit she was renamed VICTORIA, passing in 1898 to North American Mail S.S. Co. of Tacoma. She passed through various owners and spent almost all of the remainder of her career running to Alaska. In 1954 she was reported sold for conversion into a log barge, being renamed STRAITS No. 27, and after a career of some 86 years, she was sold in 1956 to Japanese shipbreakers.

DEMERARA (1872-1887) Iron Steamship.
1904 gross, 1231 net. 307 x 34 x 24 feet. Passengers: 46 1st class.
2-Cylinder Compound Inverted steam engine manufactured by the shipbuilder.
Speed 10½ knots.

Launched 1872 by J. and G. Thomson, Glasgow, for the company's Glasgow-West Indies service. Sailed from Liverpool 25.12.1887 for Gibraltar and went missing with all hands.
Sistership: TRINIDAD.

SCYTHIA (1874-1899) Iron Steamship.
4557 gross, 2907 net. 420 x 42 x 34 feet. Passengers: 300 1st class, 1100 2nd class. 2-Cylinder Compound Inverted steam engine manufactured by the shipbuilder. Speed 13 knots.

Launched 28.10.1874 by J. and G. Thomson, Glasgow, for the company's North Atlantic services. In 1899 she was sold to Italian shipbreakers.
Sistership: BOTHNIA.

GALLIA (1878-1897) Iron Steamship.
4809 gross, 2898 net. 430 x 44 x 34 feet. Passengers: 300 1st class, 1200 3rd class. 3-Cylinder Compound Inverted steam engine manufactured by the shipbuilder. Speed 15 knots.

Launched 12.11.1878 by J. and G. Thomson, Glasgow, for the company's North Atlantic services. In 1896 she was chartered to the Spanish Government for a series of trooping voyages to Cuba, where a revolt had broken out. She was transferred to the Spanish flag and renamed DON ALVARO DE BAZAN. In 1897 she was returned to Cunard, who sold her 17.11.1897 to Beaver Line Associated Steamers Ltd. She was then employed carrying emigrants to Canada, but was sold in 1899 to Allan Line of Glasgow. Almost immediately, she stranded 18.5.1899 at Point Sorel, near Quebec, and although refloated, the cost of repairs was found to be prohibitive. Sold to French shipbreakers, she was broken up in 1900 at Cherbourg.

SERVIA (1881-1902) Iron Steamship.
7392 gross, 3971 net. 515 x 52 x 37 feet. Passengers: 480 1st class, 750 3rd class. 3-Cylinder Compound Inverted steam engine manufactured by the shipbuilder. Speed 16 knots.

Launched 1.3.1881 by J. and G. Thomson, Glasgow, for the company's North Atlantic services. She was originally to have been named SAHARA, but this was changed prior to her launch. On 7.6.1893 she collided with the American sailing vessel A. McCALLUM which sank shortly afterwards with the loss of one of her crew. In 1899 she was taken up for service as a transport during the Boer War; work for which she was not particularly well suited. In 6.1900 she resumed North Atlantic service, but proved to be expensive and was soon withdrawn and in 1902 was sold to Thos. Ward for demolition at Preston.

CATALONIA (1881-1901) Iron Steamship.
4841 gross, 3093 net. 429 x 43 x 33 feet. Passengers: 200 1st class, 1500 3rd class. 2-Cylinder Compound Inverted steam engine manufactured by the shipbuilder. Speed 13 knots.

Launched 14.5.1881 by J. and G. Thomson, Glasgow, for the company's North Atlantic services. In 1899 she was taken up by the British Government for service as a transport during the Boer War. In 1901 she was sold to shipbreakers.

CEPHALONIA (1882-1900) Iron Steamship.
5517 gross, 3490 net. 430 x 46 x 34 feet. Passengers: 200 1st class, 1500 3rd class. 2-Cylinder Compound Inverted steam engine manufactured by the shipbuilder. Speed 14 knots.

Launched 20.5.1882 by Laird Brothers, Birkenhead, for the company's North Atlantic services. In 1899 she was taken up by the British Government for service as a transport during the Boer War. In 1900 she was sold to the Chinese Eastern Railway Co. Ltd. of St. Petersburg and renamed HAI LOR. Her end came in 1904 when she was sunk as a blockship at Port Arthur.

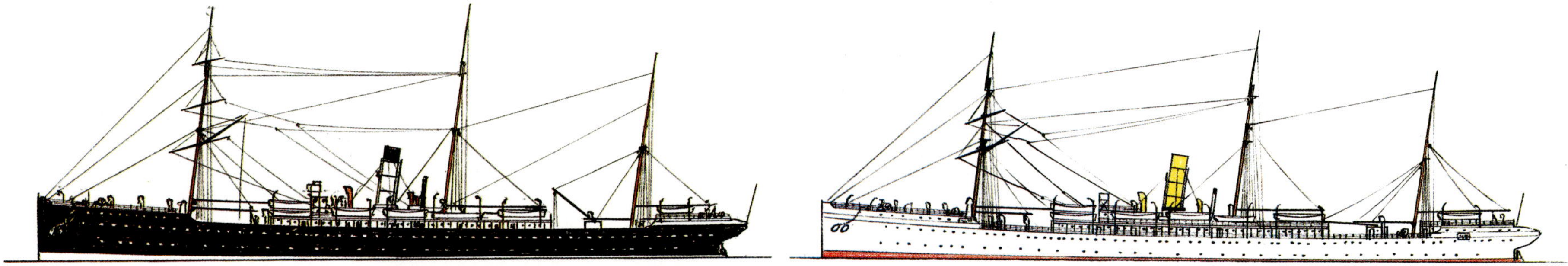

PAVONIA (1882-1900) Iron Steamship.
5588 gross, 3490 net. 430 x 46 x 34 feet. Passengers: 200 1st class, 1500 3rd class. 2-Cylinder Compound Inverted steam engine manufactured by the shipbuilder. Speed 14 knots.

Launched 3.6.1882 by J. and G. Thomson, Glasgow, for the company's North Atlantic services. In 1899 she was taken up by the British Government for service as a transport during the Boer War. Having being built for colder climates and shorter voyages, storing for large numbers of troops created considerable difficulties. After Government service she was withdrawn from service in 1900 and sold to the breakers.

AURANIA (1882-1905) Iron Steamship.
7269 gross, 4030 net. 470 x 57 x 37 feet. Passengers: 480 1st class, 700 3rd class.
2-Cylinder Compound Inverted steam engine manufactured by the shipbuilder. Speed 16 knots.

Launched 26.12.1882 by J. and G. Thomson, Glasgow, for the company's North Atlantic services. She experienced a disastrous maiden voyage as a result of severe engine damage, and was towed in to New York 4.7.1883 by a tug. There followed a lengthy refit at her builders and she did not re-enter service until 4.1884. In 1899 she was taken up by the British Government for service as a transport during the Boer War, and unlike other North Atlantic liners on the service, she proved to be exceedingly popular. In 11.1903 she inaugurated the Trieste-New York service, and in 2.1905 she was sold to Italian shipbreakers at Genoa.

OREGON (1884-1886) Iron Steamship.
7375 gross, 3529 net. 501 x 54 x 38 feet. Passengers: 340 1st class, 92 Intermediate, 110 3rd class.
2-Cylinder Compound Inverted steam engine manufactured by the shipbuilder. Speed 18 knots.

Launched 23.6.1883 by John Elder and Co., Glasgow, for Guion and Co., Liverpool. As a result of financial difficulties, the shipbuilder soon repossessed the vessel and in 5.1884 she was purchased by Cunard for their North Atlantic services. As a result of the threat of war with Russia, she was requisitioned by the Admiralty for service as a scouting cruiser and 6 x 64-pdr. muzzle-loading smoothbore guns were fitted. Her naval career was short and in 7.1885 she was returned to Cunard, £55,000 having been spent on her conversion. On 14.3.1886 she was rammed by an unknown schooner off Fire Island in New York Bay. The blow proved to be fatal as the watertight doors could not be properly closed because of coal dust. All passengers and crew were taken off by the German steamer FULDA which was in the vicinity.

UMBRIA (1884-1910) Steel Steamship.
7718 gross, 3268 net. 501 x 57 x 38 feet. Passengers: 550 1st class, 800 3rd class. 2-Cylinder Compound Inverted steam engine manufactured by the shipbuilder. Speed 19 knots.

Launched 26.6.1884 by John Elder and Co., Glasgow, for the company's North Atlantic services. On 16.4.1890 she rescued the crew of the barque MAGDALENA in the Atlantic. In 1900 she was taken up by the British Government for a service as a transport during the Boer War. In 1903 there was a mysterious attempt to blow her up in New York, a case full of dynamite being shipped, complete with electric detonators. By 1910 she had been outclassed by newer ships in the fleet and was sold to Forth Shipbreaking Co. for about £20,000.
Sistership: ETRURIA.

CAMPANIA (1892-1914) Steel Steamship.
12950 gross, 4974 net. 601 x 65 x 37 feet. Passengers: 600 1st class, 400 2nd class, 1000 3rd class. Two 5-Cylinder Triple Expansion steam engines manufactured by the shipbuilder, driving twin screws. Speed 21 knots.

Launched 8.9.1892 by Fairfield Shipbuilding and Engineering Co. Ltd., Glasgow, for the company's North Atlantic services. In 1893 she secured the Blue Riband for record voyages both east and westbound at average speeds of 21.9 and 21.49 knots respectively. In 10.1914 she was sold to Thos. Ward for demolition, but the Admiralty considered her worthy of conversion and purchsed the hull on 27.11.1914 and sent her to Cammell Laird and Co. at Birkenhead. There she was converted to an experimental aircraft-carrier for operation with the Grand Fleet. Various changes were made during her naval career, and the final design was not completed until spring 1916. She was lost on 5.11.1918 in the Firth of Forth when she dragged her anchors during a severe gale and was swept onto the ram of HMS REVENGE. She remained afloat for some hours but eventually settled on the bottom. Her crew was saved.
Sistership: LUCANIA.

CAMPANIA as an aircraft carrier in 1915

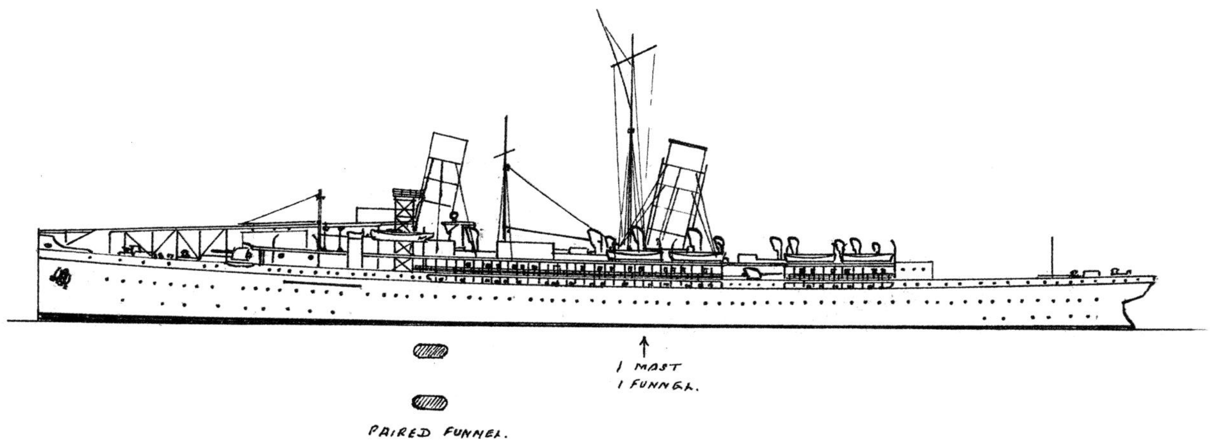

CAMPANIA in her final rig as an aircraft carrier

SYLVANIA (1895-1910) Steel Steamship.
5598 gross, 3623 net. 445 x 49 x 31 feet.
Two 3-Cylinder Triple Expansion steam engines manufactured by the shipbuilder, driving twin screws. Speed 13 knots.

Launched 24.1.1895 by London and Glasgow Shipbuilding and Engineering Co. Ltd., Glasgow, for the company's North Atlantic cargo services. In 1910 she was sold to Italian shipbreakers.
Sistership: CARINTHIA.

CARINTHIA (1895-1900) Steel Steamship.
5598 gross, 3623 net. 445 x 49 x 31 feet.
Two 3-Cylinder Triple Expansion steam engines manufactured by the shipbuilder, driving twin screws. Speed 13 knots.

Launched 24.5.1895 by London and Glasgow Shipbuilding and Engineering Co. Ltd., Glasgow, for the company's North Atlantic cargo services. In 1899 she was chartered by the British Government to carry mules from New Orleans to South Africa, and was wrecked 15.5.1900 at Gravois Point, Hayti, when on passage to Table Bay.
Sistership: SYLVANIA.

PAVIA (1897-1928) Steel Steamship.
2945 gross, 1885 net. 332 x 45 x 15 feet.
3-Cylinder Triple Expansion steam engine manufactured by Barclay, Curle and Co. Ltd., Glasgow. Speed $10\frac{1}{2}$ knots.

Launched 1897 by Workman, Clark and Co. Ltd., Belfast, for the company's Mediterranean cargo service. In 1928 she was sold to the shipbreakers.
Sisterships: CYPRIA and TYRIA.

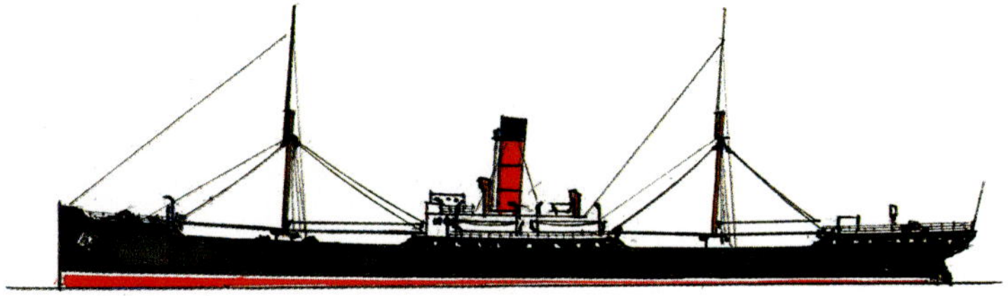

VERIA (1899-1915) Steel Steamship.
3229 gross, 2064 net. 330 x 45 x 16 feet.
3-Cylinder Triple Expansion steam engine manufactured by Wallsend Slipway Co. Ltd., Newcastle. Speed $10\frac{1}{2}$ knots.

Launched 1899 by Armstrong Whitworth and Co. Ltd., Newcastle, for the company's Mediterranean cargo service. On 7.12.1915 she was captured by a German submarine 24 miles N.W. by W. from Alexandria, and sunk by bombs. There were no casualties.
Sistership: BRESCIA.

As completed

ULTONIA (1898-1917) Steel Steamship.
8845 gross, 5748 net. 500 x 57 x 33 feet.
Two 3-Cylinder Triple Expansion steam engines manufactured by Sir C. Furness, Westgarth and Co. Ltd., Middlesbrough, driving twin screws. Speed 13 knots.

Launched 4.6.1898 by C. S. Swan and Hunter Ltd., Newcastle, for the company's North Atlantic cargo services. She had been laid down for Sir Christopher Furness but was purchased by Cunard on the stocks. In 1899 she was fitted with accommodation for 675 3rd class passengers, and this was further increased to 2100 in 1904. At that time she was transferred to the Fiume — New York service. On 27.6.1917 she was torpedoed and sunk by a German submarine 190 miles S.W. from Fastnet. One crew member was killed.

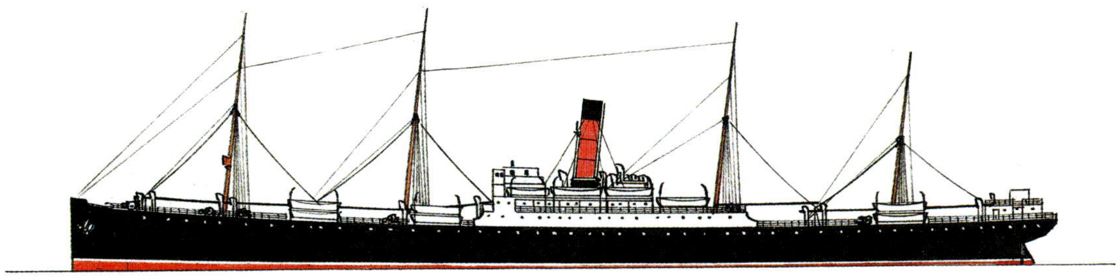

As a passenger ship

SAXONIA (1899-1925) Steel Steamship.
14281 gross, 9100 net. 580 x 64 x 38 feet. Passengers: 164 1st class, 200 2nd class, 1600 3rd class. Two 4-Cylinder Quadruple Expansion steam engines manufactured by the shipbuilder, driving twin screws. Speed 15 knots.

Launched 16.12.1899 by John Brown and Co. Ltd., Glasgow, for the company's North Atlantic services. In 1914 she was taken up by the British Government for service as a troopship, and for six months was used to accommodate German prisoners of war at London. After the hostilities she was extensively refitted at Tilbury in 1920, during which time the height of her funnel was reduced by 15 feet. After another five years service she was sold in 3.1925 for £47,000 to Dutch shipbreakers.
Sistership: IVERNIA.

With shortened funnel after her post-war refit

IVERNIA (1899-1917) Steel Steamship.
14058 gross, 9052 net. 582 x 64 x 37 feet. Passengers: 164 1st class, 200 2nd class, 1600 3rd class. Two 4-Cylinder Quadruple Expansion steam engines manufactured by Wallsend Slipway Co. Ltd., Newcastle, driving twin screws. Speed 15 knots.

Launched 21.9.1899 by C. S. Swan and Hunter Ltd., Newcastle, for the company's North Atlantic services. On 1.1.1917 she was torpedoed and sunk by a German submarine near Cape Matapan, Greece. At the time of the attack, she was carrying troops to Alexandria and 3 Officers and 82 other ranks were lost. In addition 36 crew were drowned.
Sistership: SAXONIA.

CARPATHIA (1902-1918) Steel Steamship.
13555 gross, 8660 net. 540 x 64 x 37 feet. Passengers: 204 2nd class, 1500 3rd class. Two 4-Cylinder Quadruple Expansion steam engines manufactured by Wallsend Slipway Co. Ltd., Newcastle, driving twin screws. Speed 14 knots.

Launched 6.8.1902 by C. S. Swan and Hunter Ltd., Newcastle, for the company's North Atlantic services. On 15.4.1912 when on passage from New York to the Mediterranean, she picked up a distress call from the steamer TITANIC. She immediately altered course, and was able to rescue 705 people. At the commencement of the war she was not taken up for service as a troopship, but continued on the North Atlantic service. On 17.7.1918 she was torpedoed and sunk by a German submarine about 170 miles from Bishop Rock. Five of her crew were lost.

YAMUNA

SLAVONIA (1904-1909) Steel Steamship.
10606 gross, 6725 net. 510 x 59 x 22 feet. Passengers: 71 1st class, 74 2nd class, 1954 3rd class. Two 3-Cylinder Triple Expansion steam engines manufactured by Wallsend Slipway Co. Ltd., Newcastle, driving twin screws. Speed 13 knots.

Launched 15.11.1902 by Sir J. Laing and Sons Ltd., Sunderland, as YAMUNA for the British India Steam Navigation Co. Ltd., London. On 30.1.1904 she was purchased by the company for the Trieste-New York service and renamed SLAVONIA. Stranded on 11.6.1909 on the island of Flores in the Azores, her passengers and crew were rescued.

PANNONIA (1902-1922) Steel Steamship.
9851 gross, 6210 net. 486 x 59 x 33 feet. Passengers: 40 1st class, 800 2nd class.
Two 3-Cylinder Triple Expansion steam engines manufactured by the shipbuilder, driving twin screws. Speed 13 knots.

Launched 5.9.1902 by John Brown and Co. Ltd., Glasgow, for the company's Trieste-New York service. She had been laid down for Sir Christopher Furness, but was purchased by Cunard on the stocks. Sold 10.1922 to German shipbreakers for demolition at Hamburg.

BRESCIA (1903-1930) Steel Steamship.
3235 gross, 2058 net. 350 x 45 x 16 feet.
3-Cylinder Triple Expansion steam engine manufactured by Wallsend Slipway C Ltd., Newcastle. Speed 10½ knots.

Launched 1903 by J. L. Thompson and Sons Ltd., Sunderland, for the company's Mediterranean cargo service. Sold in 1930 to shipbreakers.
Sistership: VERIA.

CARONIA (1904-1932) Steel Steamship.
19687 gross, 10306 net. 650 x 72 x 40 feet. Passengers: 300 1st class, 350 2nd class, 2000 3rd class. Two Quadruple Expansion 4-Cylinder steam engines manufactured by the shipbuilder, driving twin screws. Speed 18 knots.

Launched 13.7.1904 by John Brown and Co. Ltd., Glasgow, for the company's North Atlantic services. Requisitioned by the Admiralty in 8.1914 she was converted to an auxiliary cruiser and fitted with 8 x 4.7'' guns which had been taken from outdated cruisers. She finished the war without serious incident and in 1924 was given an extensive refit at Barrow prior to a term on the Canada service. She was laid up at Sheerness 1931 and in 1.1932 was sold to shipbreakers at Blyth. Before demolition could commence she was resold to Japanese shipbreakers and was renamed TAISEIYO MARU for the voyage to Osaka.
Sistership: CARMANIA.

CARONIA following her 1924 refit

CARMANIA (1905-1932) Steel Steamship.
19524 gross, 9982 net. 650 x 72 x 40 feet. Passengers: 300 1st class, 350 2nd class, 2000 3rd class. Three Steam Turbines manufactured by the shipbuilder, driving triple screws. Speed 18 knots.

Launched 21.2.1905 by John Brown and Co. Ltd., Glasgow, for the company's North Atlantic services. Requisitioned by the Admiralty in 8.1914 she was converted to an auxiliary cruiser and fitted with 8 x 4.7'' guns which had been take from outdated cruisers. On 14.9.1914 she surprised the German auxiliary cruiser CAP TRAFALGAR which was bunkering off Trinidad Island. An intense artillery duel commenced, which resulted in the loss of CAP TRAFALGAR and heavy damage to CARMANIA. Following repairs at Gibraltar, she was returned to service and in 5.1916 was returned to Cunard. In 1923 she went back to her builders for an extensive refit and then resumed her career on the Canadian service. The shipping slump of the early 1930s brought her career to an end and she was sold in 1932 for £20,000 to Hughes Bolckow for demolition at Blyth.
Sistership: CARONIA.

CARMANIA following her 1923 refit

LUSITANIA (1906-1915) Steel Steamship.
31550 gross, 9145 net. 762 x 87 x 56 feet. Passengers: 563 1st class, 464 2nd class, 1138 3rd class. Four Steam Turbines manufactured by the shipbuilder, driving quadruple screws. Speed 25 knots.

Launched 7.6.1906 by John Brown and Co. Ltd., Glasgow, for the company's North Atlantic services. She gained the Blue Riband in 10.1907 with an average speed of 23.99 knots between Queenstown and the Ambrose Light, but MAURETANIA soon proved to be the faster of the two sisters. Following the outbreak of war, despite German warnings about the war zone around the British Isles, she continued on the North Atlantic service. On 7.5.1915, when off the Old Head of Kinsale, she was torpedoed and sunk by the German submarine U20. She sank within 20 minutes and 1198 passengers and crew were lost. In Britain the news was received with horror and in Germany with acclamation, since it had been reported there that she was carrying ammunition and troops. This was one of the first examples of total war.
Sistership: MAURETANIA.

MAURETANIA (1906-1935) Steel Steamship.
31938 gross, 8948 net. 762 x 88 x 57 feet. Passengers: 563 1st class, 464 2nd class, 1138 3rd class. Four Steam Turbines manufactured by Wallsend Slipway Co. Ltd., Newcastle, driving quadruple screws. Speed 25 knots.

Launched 20.9. 1906 by Swan, Hunter and Wigham Richardson Ltd., Newcastle, for the company's North Atlantic services. She quickly secured the record for the Atlantic crossing, and when in 9.1909 she covered the distance between Daunts Rock and Sandy Hook at an average speed of 26.06 knots, she set a record unbeaten for 20 years. In 1915 she was taken up by the British Government for service as a troopship and served in the Gallipoli campaign. She then spent some time as a hospital ship, before returning once again to trooping. Returned to her owners in 1919, she suffered extensive fire damage 25.7.1921 at Southampton. It was decided to return her to her builders where she was extensively refitted and rebuilt. On her return to service she was employed cruising off-season and in 1930 was repainted with a white hull. She was finally sold to shipbreakers in 1935 and broken up at Rosyth by Metal Industries Ltd.
Sistership: LUSITANIA.

MAURETANIA as a hospital ship

MAURETANIA following her post-war refit

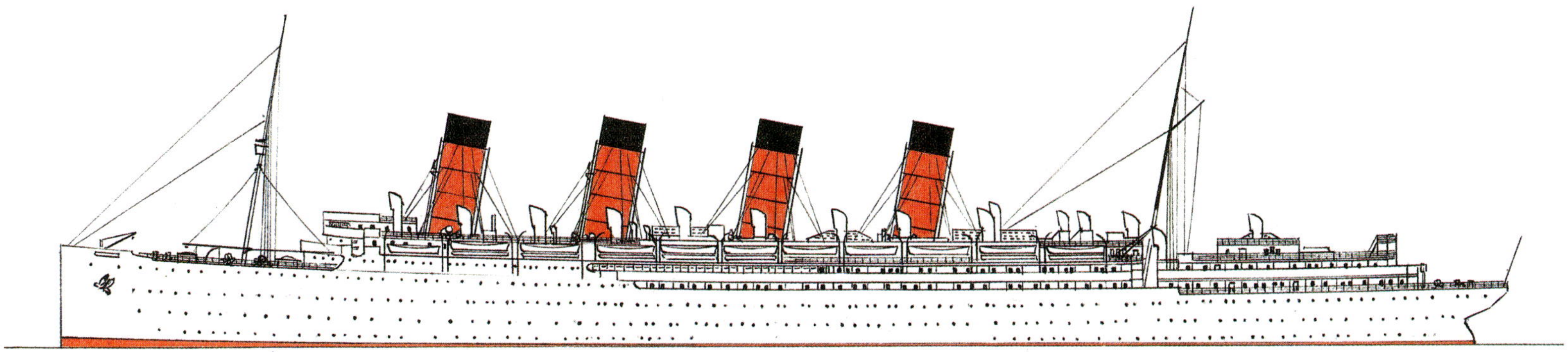

MAURETANIA in cruising livery in the 1930's

FRANCONIA (1910-1916) Steel Steamship.
18150 gross, 11247 net. 600 x 71 x 40 feet. Passengers: 300 1st class, 350 2nd class, 2200 3rd class. Two Quadruple Expansion 4-Cylinder steam engines manufactured by Wallsend Slipway Co. Ltd., Newcastle, driving twin screws. Speed 17 knots.

Launched 23.7.1910 by Swan, Hunter and Wigham Richardson Ltd., Newcastle, for the company's North Atlantic services. In 1915 she was taken up by the British Government for service as a troop transport and 4.10.1916 was torpedoed and sunk by a German submarine 195 miles SE of Malta. She sank within 50 minutes and 12 of her crew were lost.
Sistership: LACONIA.

CONSUELO in the livery of T. Wilson, Sons and Co. Ltd.

ALBANIA

ALBANIA (1911-1912) Steel Steamship.
7640 gross, 3960 net. 461 x 52 x 31 feet. Passengers 50 2nd class, 800 3rd class. Two Triple Expansion 3-Cylinder steam engines manufactured by T. Richardson and Sons Ltd., Hartlepool, driving twin screws. Speed 11 knots.

Launched 3.2.1900 by C. S. Swan and Hunter Ltd., Newcastle, as CONSUELO for T. Wilson, Sons and Co. Ltd., Hull. In 4.1908 she was sold to the Cairn Line of Steamships Ltd., Newcastle, for their Thomson Line service to Canada and was renamed CAIRNRONA. She was purchased 3.1911 by Cunard for their new Canada service, and commenced her first voyage in their colours as ALBANIA. As she was far below the standard required for the route she was soon sold, and in 1912 went to Andrew Weir and Co. for £21,000. They renamed her POLERIC and used her for tramping services. In 1927 she was laid up at Shanghai and soon sold for £17,000 to Japanese shipbreakers.

AUSONIA as built

AUSONIA following her refit in 1913

AUSONIA (1911-1918) Steel Steamship.
7907 gross, 4955 net. 450 x 54 x 29 feet. Passengers: 90 2nd class, 1000 3rd class. Two Triple Expansion 3-Cylinder steam engines manufactured by Palmers' Shipbuilding and Iron Co. Ltd., Jarrow, driving twin screws. Speed 12 knots.

Launched 18.8.1909 by Swan, Hunter and Wigham Richardson Ltd., Newcastle, as TORTONA for the Thomson Line service operated by the Cairn Line of Steamships Ltd., Newcastle. She was purchased 3.1911 by Cunard for their new Canada service, and commenced her first voyage in their colours as AUSONIA. On 30.5.1918 she was torpedoed by a German submarine about 600 miles W. of the Irish coast. The submarine surfaced and sank her with gunfire. A total of 44 passengers and crew were lost.

ASCANIA (1911-1918) Steel Steamship.
9111 gross, 5699 net. 466 x 56 x 29 feet. Passengers: 200 2nd class, 1500 3rd class. Two Triple Expansion 3-Cylinder steam engines manufactured by Palmers' Shipbuilding and Iron Co. Ltd., Jarrow, driving twin screws. Speed 13 knots.

Launched 6.3.1911 by Swan, Hunter and Wigham Richardson Ltd., Newcastle, for the company's new Canada service. She was purchased on the stocks, having been laid down as GERONA for Thomson Line. She was wrecked on 13.6.1918, 20 miles E. of Cape Ray, Newfoundland, when on a voyage from Liverpool to Montreal in ballast.

LACONIA (1911-1917) Steel Steamship.
18099 gross, 11226 net. 600 x 71 x 40 feet. Passengers: 300 1st class, 350 2nd class, 2200 3rd class.
Two Quadruple Expansion 4-Cylinder steam engines manufactured by Wallsend Slipway Co. Ltd., Newcastle, driving twin screws. Speed 17 knots.

Launched 27.7.1911 by Swan, Hunter and Wigham Richardson Ltd., Newcastle, for the company's North Atlantic services. Requisitioned by the Admiralty in 10.1914, she was converted to an armed merchant cruiser. In 9.1916 she was returned to her owners and was torpedoed and sunk 25.2.1917 when 160 miles off Fastnet. The attack took place at night and 12 lives were lost.
Sistership: FRANCONIA.

ANDANIA (1913-1918) Steel Steamship.
13405 gross, 8275 net. 520 x 64 x 43 feet. Passengers: 520 2nd class, 1540 3rd class.
Two Quadruple Expansion 4-Cylinder steam engines manufactured by the shipbuilder, driving twin screws. Speed 15 knots.

Launched 22.3.1913 by Scotts' Shipbuilding and Engineering Co. Ltd., Greenock, for the company's Canada service. On 27.1.1918 she was torpedoed and sunk by a German submarine off Rathlin Island. Seven crew were lost in the ensuing explosion.
Sistership: ALAUNIA.

ALAUNIA (1913-1916) Steel Steamship.
13405 gross, 8261 net. 520 x 64 x 43 feet. Passengers: 520 2nd class, 1540 3rd class.
Two Quadruple Expansion 4-Cylinder steam engines manufactured by the shipbuilder, driving twin screws. Speed 15 knots.

Launched 9.6.1913 by Scotts' Shipbuilding and Engineering Co. Ltd., Greenock, for the company's Canada service. On 19.10.1916 she struck a mine and sank in the English Channel, 2 miles S. of the Royal Sovereign lightship. Two crew were lost.
Sistership: ANDANIA.

AURANIA (1916-1918) Steel Steamship.
13936 gross, 8499 net. 506 x 65 x 42 feet. Passengers: 506 1st class, 1530 3rd class.
Four Steam Turbines manufactured by Wallsend Slipway Co. Ltd., Newcastle, single reduction geared to twin screws. Speed 15 knots.

Launched 16.7.1916 by Swan, Hunter and Wigham Richardson Ltd., Newcastle. She had been designed for the Canada service, but on account of the war, was taken up for service as a troop transport on the North Atlantic. Painted grey all over, she almost certainly never sailed in Cunard colours. On 4.2.1918 she was torpedoed by a German submarine about 40 miles N. of Londonderry. She was taken in tow, but the towing wire carried away and she was wrecked on submerged rocks off Tobermory. Nine crew were lost.

AQUITANIA (1913-1950) Steel Steamship.
45647 gross, 21993 net. 868 x 97 x 49 feet. Passengers: 597 1st class, 614 2nd class, 2052 3rd class. Four Steam Turbines manufactured by the shipbuilder, driving quadruple screws. Speed 23 knots.

Launched 21.4.1913 by John Brown and Co. Ltd., Glasgow, for the company's North Atlantic services. On 2.8.1914 she was hired by the Admiralty for service as an armed merchant cruiser. Painted grey, she was fitted with 6'' guns and on 8.8.1914 commenced her first Atlantic patrol. On 22.8.1914 she was in collision with the Leyland liner CANADIAN and her naval career was at an end. By late September she had been decommissioned and was laid up until taken up as a troop transport. She then served periods as both a hospital ship and a transport, and at one stage was the recipient of dazzle painting. Late in 1919 she was sent to Swan, Hunter on the Tyne for an extensive refit and engine overhaul. Over the following years, her passenger accommodation was substantially altered to take account of developments and improvements on the North Atlantic. The depression of the 1930s resulted in a dramatic drop in passenger numbers and she was consequently employed on occasional cruising.With the outbreak of the Second War, she was once again employed on trooping; a role she retained until 1948. For a short period she was employed on the Canadian service, but in 1950 she was withdrawn from service and sold to shipbreakers at Faslane.

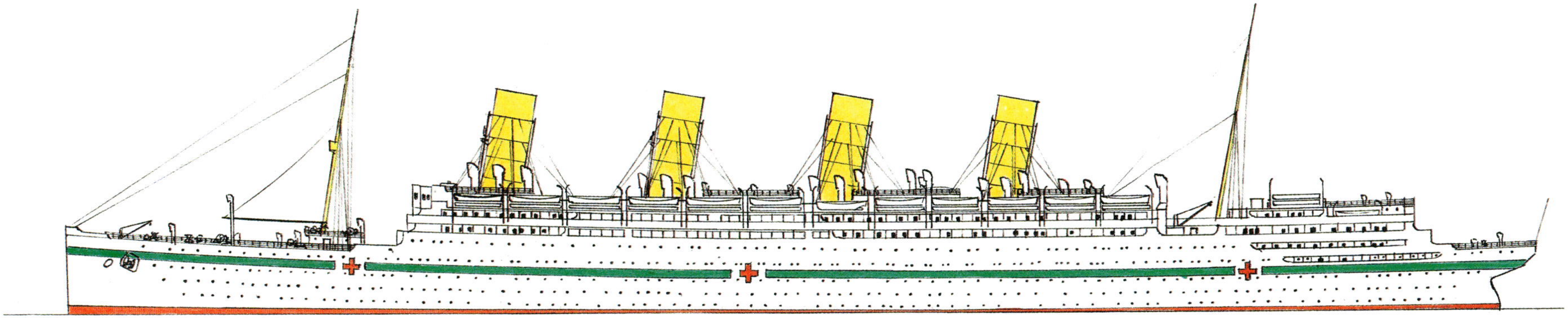

AQUITANIA as a hospital ship

AQUITANIA as a troopship in war-time camouflage

ROYAL GEORGE (1916-1922) Steel Steamship.
11146 gross, 5685 net. 525 x 60 x 27 feet. Passengers: 710 1st class, 290 2nd class.
Three Steam Turbines manufactured by the shipbuilder, driving triple screws. Speed 19 knots.

Launched 28.5.1907 by Fairfield Co. Ltd., Glasgow, as HELIOPOLIS for the Egyptian Mail Steamship Co., London. She was intended for their new express mail service between Marseilles and Alexandria, but only served a few months before being laid up at Marseilles. Early in 1910 she was sold to Canadian Northern Steamships of Toronto and renamed ROYAL GEORGE. They placed her on a North Atlantic service between Avonmouth and Canada. In 1914 she was taken up by the British Government for service as a troopship, and in 6.1916 the company and its ships were purchased by Cunard. In 1920 she was sent to Cherbourg to act as an accommodation ship for emigrants. In a very short time she had been virtually gutted and all expensive fittings removed. Towed back to the Fal, she was laid up for a period prior to her sale in 1922 to German shipbreakers.

VALACIA (1916-1931) Steel Steamship.
6526 gross, 4100 net. 460 x 57 x 28 feet.
Triple Expansion 3-Cylinder steam engine manufactured by Rankin and Blackmore, Greenock. Speed 13 knots.

Launched 1910 by Russell and Co., Port Glasgow, as LUCERIC for Bank Line Ltd., Glasgow. Purchased in 1916 for the company's North Atlantic cargo services, she was renamed VALACIA. In 1931 she was sold to Italian owners and renamed ERNANI. Taken over by German forces in 1940, she was renamed SLEIPNER II, and camouflaged as the Dutch steamer ENGGANO for an attempt at breaking the Allied blockade of Bordeaux. On 28.6.1941 she was torpedoed and sunk by a German submarine whilst making the attempt.

KURSK (1918-1920) Steel Steamship.
7869 gross, 4637 net. 450 x 56 x 31 feet.
Two Quadruple Expansion 4-Cylinder steam engines manufactured by the shipbuilder, driving twin screws. Speed 16 knots.

Launched 1910 by Barclay, Curle and Co. Ltd., Glasgow, for the Russian East Asiatic S.S. Co. Ltd., Libau. Following the Russian Revolution, she came under the control of The Shipping Controller, London on 21.2.1918, and management was allocated to Cunard who employed her almost exclusively on trooping. On 17.6.1920 she was returned to the Danish East Asiatic Co, and renamed POLONIA. Sold to Polish owners in 1930, she was not renamed, but sailed under this name until broken up in 1939.

CZAR (1918-1920) Steel Steamship.
6516 gross, 3812 net. 425 x 53 x 29 feet.
Two Quadruple Expansion 4-Cylinder steam engines manufactured by the shipbuilder, driving twin screws. Speed 16 knots.

Launched 1912 by Barclay, Curle and Co. Ltd., Glasgow, for the Russian East Asiatic S.S. Co. Ltd., Libau. Following the Russian Revolution, she came under the control of The Shipping Controller, London on 8.3.1918, and management was first allocated to Ellerman's Wilson Line. After a few months it was transferred to Cunard who employed her almost exclusively on trooping. On 8.8.1920 she was returned to the Danish East Asiatic Co., and after a refit resumed service as ESTONIA. She passed to Polish owners in 1930 and was renamed PULASKI, serving under this name throughout the Second World War. In 1946 she passed to the Ministry of Transport and was renamed EMPIRE PENRYN. After three years she was declared surplus to requirements and in 1949 was broken up at Blyth.

CZARITZA (1918-1920) Steel Steamship.
6598 gross, 4119 net. 440 x 53 x 29 feet.
Two Quadruple Expansion 4-Cylinder steam engines manufactured by the shipbuilder, driving twin screws. Speed 16 knots.

Launched 1915 by Barclay, Curle and Co. Ltd., Glasgow, for the Russian East Asiatic S.S. Co. Ltd., Libau. Following the Russian Revolution, she came under the control of The Shipping Controller, London on 16.3.1918, and management was first allocated to Ellerman's Wilson Line. After a few months it was transferred to Cunard who employed her almost exclusively on trooping. On 23.7.1920 she was returned to the Danish East Asiatic Co., and after a refit resumed service as LITUANIA. She too passed to Polish owners in 1930 and was renamed KOSCIUSZCO, and later GDYNIA, and when under the British flag, EMPIRE HELFORD. In 1950 she was sent to the breakers.

VENUSIA (1919-1923) Steel Steamship.
5222 gross, 3172 net. 400 x 52 x 28 feet.
Triple Expansion 3-Cylinder steam engine manufactured by the shipbuilder. Speed $11\frac{1}{2}$ knots.

Launched 22.8.1918 by Harland and Wolff Ltd., Belfast, as WAR SNAKE for The Shipping Controller, London. Purchased 27.1.1919 for the company's North Atlantic cargo service, she was renamed VENUSIA. In 1923 she was sold to S. and J. Thompson, London, and renamed RIVER DELAWARE, and in 1931 passed to Italian owners as RINA CORRADO. On 9.11.1941 she was shelled and sunk 120 miles S. E. of Punta Stilo, Calabria, when on passage from Messina to Tripoli with munitions.

VERENTIA (1919-1926) Steel Steamship.
5185 gross, 3152 net. 400 x 52 x 28 feet.
Triple Expansion 3-Cylinder steam engine manufactured by the shipbuilder. Speed 11½ knots.

Launched 9.5.1918 by Harland and Wolff Ltd., Belfast, as WAR LEMUR for The Shipping Controller, London. Purchased 27.1.1919 for the company's North Atlantic cargo service, she was renamed VERENTIA. In 1926 she was sold to A. Weir and Co., Glasgow, and renamed FORERIC; sold again in 1927 to become GALVAN, and in 1937 PEI TAI, she was finally sold to Japanese owners in 1938 as HOKUTAI MARU. On 30.3.1944 she was sunk by American carrier-based aircraft off Babelthuap in the Pacific Ocean.

VITELLIA (1919) Steel Steamship.
4449 gross, 2681 net. 375 x 51 x 26 feet.
Triple Expansion 3-Cylinder steam engine manufactured by the shipbuilder. Speed 11½ knots.

Launched 1918 by Earle's Co. Ltd., Hull, as WAR PINTAIL for The Shipping Controller, London. Purchased 18.1.1919 for the company's North Atlantic cargo service, she was renamed VITELLIA. Her career with Cunard was short for on 5.6.1919 she was sold to the Anchor Line of Glasgow (which had been controlled since 1912 by Cunard), who continued to trade her on the North Atlantic, but without a change of name. On 13.12.1923 she was sold again, this time to Scindia Steam Navigation Co. of Bombay, and was renamed JALARASHMI. After 30 years of trading she was sold to other Indian owners in 1953 as ASHA, and was finally broken up in 1955 in Japan.

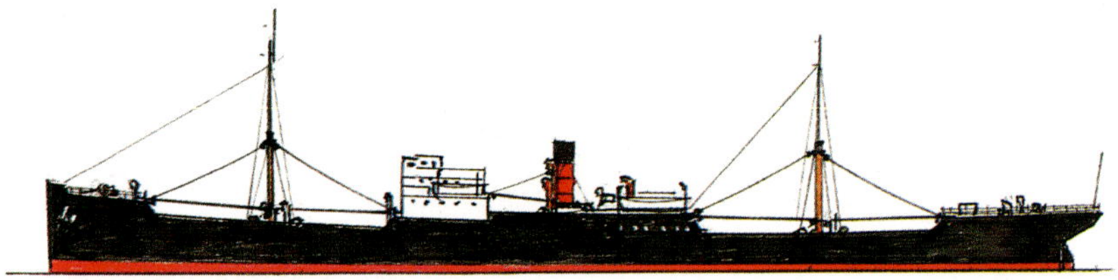

VELLAVIA (1919-1931) Steel Steamship.
5272 gross, 3195 net. 400 x 52 x 28 feet.
Triple Expansion 3-Cylinder steam engine manufactured by George Clark Ltd., Sunderland. Speed 11½ knots.

Launched 1918 by Armstrong, Whitworth and Co. Ltd., Newcastle, as WAR SETTER for The Shipping Controller, London. Purchased 28.5.1919 for the company's North Atlantic service, she was renamed VELLAVIA. In 1925 she too was sold to S. and J. Thompson, London, and was renamed RIVER TIGRIS. Sold to Italian owners in 1931 she became INES CORRADO. At the outbreak of war she sheltered in Argentina, and was taken over there in 12.1941 by the Argentine Government and renamed RIO DIAMANTE. At the cessation of hostilities, she returned to her Italian owners as INES CORRADO. In 1959 she was sold to Japanese shipbreakers.

VARDULIA (1919-1929) Steel Steamship.
5691 gross, 3613 net. 423 x 56 x 28 feet.
Triple Expansion 3-Cylinder steam engine manufactured by Rankin and Blackmore, Greenock. Speed 12 knots.

Launched 1.1917 by Russell and Co., Port Glasgow, as VERDUN for Gow, Harrison and Co., Glasgow. Purchased in 6.1919 for the company's North Atlantic cargo service, she was renamed VARDULIA. In 1929 she was sold to Donaldson Line, without a change of name. She was lost 20.10.1935 when she foundered in the North Atlantic when on passage from Hartlepool to Botwood with a cargo of coal. Her complete crew was lost.

VIRGILIA (1919-1925). Steel Steamship.
5697 gross, 3619 net. 423 x 56 x 28 feet.
Triple Expansion 3-Cylinder steam engine manufactured by J. G. Kincaid and Co. Ltd., Greenock. Speed 11½ knots.

Launched 1918 by Russell and Co., Port Glasgow, for Gow, Harrison and Co., Glasgow. Purchased in 1919 for the company's North Atlantic cargo service, the name was retained. In 1925 she was sold to J. Chambers and Co., Liverpool, for their Lancashire Shipping Co. Ltd., and renamed CORBY CASTLE. They only retained her for a short period, for in 1928 she was sold to Japanese owners and renamed TATSUHA MARU. On 17.2.1944 she was sunk by American carrier-based aircraft off the Caroline Islands.

ALBANIA (1920-1930) Steel Steamship.
12768 gross, 7822 net. 523 x 64 x 43 feet. Passengers: 500 Cabin class.
Four Steam Turbines manufactured by the shipbuilder, double reduction geared to twin screws. Speed 13 knots.

Launched 17.4.1920 by Scotts' Shipbuilding and Engineering Co. Ltd., Greenock, for the company's Canada service. She had in fact been ordered in 1916, but construction was delayed on account of the war. She was an unsuccessful ship and was laid up in 1925 until sold in 1930 to Italian owners. Renamed CALIFORNIA, she was considerably altered by her new owners. In 1935 the Italian Government requisitioned her for service as a hospital ship during the Abyssinian campaign, and she served as such for a number of years. On 11.8.1941 she was torpedoed and sunk by British aircraft off Syracuse, Sicily. She was subsequently refloated and broken up.

SCYTHIA as built

SCYTHIA (1920-1958) Steel Steamship.
19730 gross, 11938 net. 600 x 73 x 40 feet. Passengers: 350 1st class, 350 3nd class, 1500 3rd class. Six Steam Turbines manufactured by the shipbuilder, double reduction geared to twin screws. Speed 16 knots.

Launched 23.3.1920 by Vickers Ltd., Barrow, for the company's North Atlantic services. As a result of the joiner's strike whilst she was fitting out, she was sent to Lorient in France for completion. In August 1939 she was taken up by the British Government for service as a troop transport and served as such throughout the war. On 23.11.1942 she was severely damaged by aerial torpedo whilst lying at Algiers, but was repaired and returned to service. Following release from Government service in 1948, she was given a lengthy refit in 1949/1950 at Clydebank and was then placed on the Canada service. In 1958 she was sold to shipbreakers and broken up on the Firth of Forth.
Sisterships: LACONIA and SAMARIA.

SCYTHIA in the 1950's

TYRRHENIA

TYRRHENIA/LANCASTRIA (1920-1940) Steel Steamship.
16243 gross, 9645 net. 552 x 70 x 38 feet. Passengers: 280 1st class, 364 2nd class, 1200 3rd class. Six Steam Turbines manufactured by the shipbuilder, double reduction geared to twin screws. Speed 15 knots.

Launched 31.5.1920 by William Beardmore and Co. Ltd., Glasgow, as TYRRHENIA for the company's Canada service. She had in fact been laid down for the associated Anchor Line's Mediterranean — New York service. Her maiden voyage was not until 6.1922 as fitting out had been much delayed by a joiner's strike. With a transfer to the New York service in 1924, her name was changed to LANCASTRIA, which proved to be a very popular choice. With the reduction of passenger numbers during the depression, she was regularly employed cruising, and at one time appeared with an all-white hull. In 1940 she was requisitioned by the British Government as a troop transport and was first used in the evacuation of troops from Norway. In 6.1940 she was sent to St. Nazaire to evacuate as many troops and evacuees as possible. On 17.6.1940 she was attacked by German aircraft outside the port and sank very quickly. Since no record had been kept of the numbers embarked, the losses can only be estimated. Semi-official records suggest 2500-3000, but others suggest as many as 4500.

LANCASTRIA

LACONIA (1921-1942) Steel Steamship.
19680 gross, 11829 net. 601 x 73 x 40 feet. Passengers: 350 1st class, 350 2nd class, 1500 3rd class. Six Steam Turbines manufactured by Wallsend Slipway Co. Ltd., Newcastle, double reduction geared to twin screws. Speed 16 knots.

Launched 9.4.1921 by Swan, Hunter and Wigham Richardson Ltd., Newcastle, for the company's North Atlantic services. Her launch had originally been planned for 11.1920 but was delayed because the slipway was blocked by the French liner MEDUANA, which had sunk after a fire. In 1939 she was taken up by the British Government for service as an armed merchant cruiser, and from 1941 a troop transport. On 12.9.1942 she was torpedoed and sunk by a German submarine in the South Atlantic. She was carrying over 2700 passengers, including 1800 Italian prisoners of war. Despite efforts by the submarine and other Axis craft to rescue survivors, 2275 persons were lost. Sisterships: SAMARIA and SCYTHIA.

AURANIA (1924-1942) Steel Steamship.
13984 gross, 8473 net. 519 x 65 x 39 feet. Passengers: 500 Cabin class, 1200 3rd class. Four Steam Turbines manufactured by Wallsend Slipway Co. Ltd., Newcastle, double reduction geared to twin screws. Speed 15 knots.

Launched 6.2.1924 by Swan, Hunter and Wigham Richardson Ltd., Newcastle, for the company's Canada service. In 9.1939 she was hired by The Admiralty for service as an armed merchant cruiser. Painted grey, she was fitted at Swan, Hunter, Wallsend, with 8 x 6'' guns and 2 x 3'' anti-aircraft guns, and thereafter served on patrol and escort duties in the Atlantic. On 9.3.1942 she was sold to The Admiralty. They were in need of a fleet repair ship and sent her back to Swan, Hunter at Wallsend for conversion. She returned to service in 1944 as H.M.S. ARTIFEX. Sold in 1961 to Italian shipbreakers. Sisterships: ALAUNIA and ASCANIA.

AURANIA as an armed merchant cruiser

AURANIA as the fleet repair ship H.M.S. ARTIFEX

BERENGARIA (1921-1939) Steel Steamship.
52226 gross, 22016 net. 883 x 98 x 57 feet. Passengers: 970 1st class, 830 2nd class, 1000 3rd class. Four Steam Turbines manufactured by the shipbuilder driving quadruple screws. Speed 22 knots.

Launched 23.5.1912 by Vulcan-Werke, Hamburg, as IMPERATOR for the Hamburg-America Line, Hamburg. In 8.1914 she was laid up at Hamburg for the duration of the war, and following the Armistice was allocated to the United States for service as a transport. When the Americans had no further use for her she was transferred to The Shipping Controller, London, and in 1920 was chartered to Cunard. The following year she was purchased outright for the North Atlantic services and renamed BERENGARIA. There followed an extensive period of reconditioning on the Tyne, including conversion to oil fuel. In 1935 she was again reconditioned and her appearance was slightly altered. On 3.3.1938 fire broke out in her passenger accommodation while in New York, and she returned to Southampton and was laid up. Repairs were considered uneconomical and she was sold to Sir John Jarvis for demolition at Jarrow. It was his intention that the work would go some way towards relieving unemployment on Tyneside. By the time war was declared she had been reduced down to her double bottoms. Work then ceased and it was not until 1946 that demolition was completed when the remains of the hull were towed to Rosyth.

BERENGARIA following her 1935 refit

FRANCONIA (1922-1956) Steel Steamship.
20175 gross, 12162 net. 601 x 73 x 40 feet. Passengers: 330 1st class, 420 2nd class, 950 3rd class. Six Steam Turbines manufactured by the shipbuilder, double reduction geared to twin screws. Speed 16 knots.

Launched 21.10.1922 by John Brown and Co. Ltd., Glasgow, for the company's North Atlantic services. In common with other of the company's liners at the time, she was frequently employed cruising, at which time her hull was painted white. In 9.1939 she was taken up by the British Government for service as a troop transport, and in 1945 served as the Headquarters for Winston Churchill at Yalta. Returned to her owners in 1948, she was sent for an extensive overhaul and resumed commercial service 6.1949. In 1956 she was sold to Thos. Ward for demolition and was broken up at Inverkeithing. Sistership: CARINTHIA.

FRANCONIA in cruising livery. CARINTHIA (page 46) was similar.

ASCANIA (1923-1957) Steel Steamship.
14013 gross, 8437 net. 520 x 65 x 39 feet. Passengers: 500 Cabin class, 1200 3rd class.
Four Steam Turbines manufactured by the shipbuilder, double reduction geared to twin screws. Speed 15 knots.

Launched 20.12.1923 by Armstrong, Whitworth and Co. Ltd., Newcastle, for the company's Canada service. Completion was much delayed on account of excessive costs and wages, which resulted in a complete halt to construction for several months. She finally sailed from London 22.5.1925 on her maiden voyage to Quebec and Montreal. On 14.12.1934 she went to the assistance of the British steamer USWORTH, which was experiencing difficulties in a severe storm in the North Atlantic. After an epic struggle the remaining crew of the USWORTH were taken off by lifeboats and were eventually taken on board ASCANIA. In 9.1939 she was hired by The Admiralty for service as an armed merchant cruiser. Painted grey, she was fitted at Birkenhead with 8 x 6'' guns and 2 x 3'' anti-aircraft guns, and thereafter served on patrol and scout duties in the Atlantic. From 1942 she served as a troop transport, and was finally released back to her owners in 1947. Late in 1949 she was sent to Alexander Stephen and Sons at Linthouse for a complete overhaul and returned 4.1950 to the Canada service. In 1957 she was broken up at Newport, Monmouthshire.
Sisterships: ALAUNIA and AURANIA.

ASCANIA in her final rig

ALAUNIA (1925-1944) Steel Steamship.
14030 gross, 8448 net. 519 x 65 x 39 feet. Passengers: 500 cabin class, 1200 3rd class.
Four Steam Turbines manufactured by the shipbuilder, double reduction geared to twin screws. Speed 15 knots.

Launched 7.2.1925 by John Brown and Co. Ltd., Glasgow, for the company's Canada service. In 8.1939 she was hired by The Admiralty for service as an armed merchant cruiser. Painted grey, she was fitted at Gibraltar with 8'' x 6'' guns and 2 x 3'' anti-aircraft guns, and thereafter served on patrol and escort duties in the Atlantic. Like her near sisters, she too was employed later as a troop transport and on 8.2.1944 was sold to The Admiralty for service as a fleet repair ship. In 9.1945 she commenced service as H.M.S. ALAUNIA and continued until sold in 1957 to shipbreakers at Blyth.
Sisterships: ASCANIA and AURANIA.

H.M.S. ALAUNIA, fleet repair ship

CARINTHIA (1925-1940) Steel Steamship.
20277 gross, 12088 net. 600 x 73 x 40 feet. Passengers: 240 1st class, 460 2nd class, 950 3rd class. Four Steam Turbines manufactured by the shipbuilder, double reduction geared to twin screws. Speed 16 knots.

Launched 24.2.1925 by Vickers Ltd., Barrow, for the company's North Atlantic service. She had been laid down as SERVIA, but the name was changed prior to her launch. In common with FRANCONIA, she was used frequently for cruising, and made a series of world cruises during which time she was based at New York. In 8.1939 she was hired by The Admiralty for service as an armed merchant cruiser. Painted grey, she was fitted with 8 x 6'' guns and 2 x 3'' anti-aircraft guns. On 6.6.1940 she was torpedoed by a German submarine off the coast of Northern Ireland. She sank some 30 hours later with the loss of four lives.
Sistership: FRANCONIA.

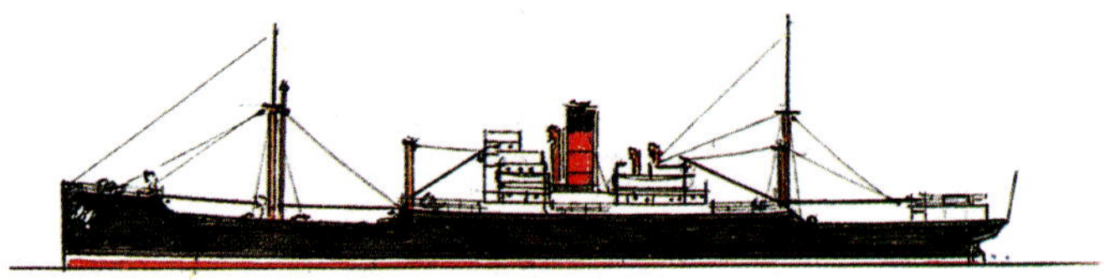

BANTRIA (1928-1953) Steel Steamship.
2402 gross, 1209 net. 292 x 45 x 20 feet.
Triple Expansion 3-Cylinder steam engine manufactured by J. Dickinson and Sons Ltd., Sunderland. Speed 12 knots.

Launched 17.5.1928 for America-Levant Line Ltd. (Stanley and John Thompson Ltd., managers), London. Cunard soon took over the Mediterranean services of this company and she was registered under the ownership of the Cunard Steam Ship Co. Ltd. In 1953 she was sold to Italian owners as GIORGINA CELLI and was sold again in 1955 to become SACRUM COR. She was eventually broken up in 1968 at Vado, Italy.
Sisterships: BOTHNIA, BACTRIA and BOSNIA.

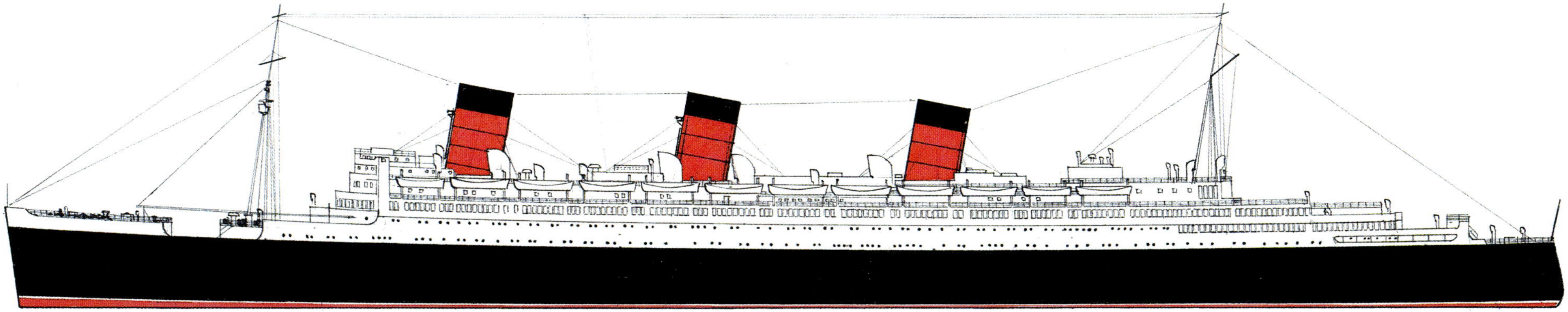

QUEEN MARY (1934-1967) Steel Steamship.
81235 gross, 34253 net. 975 x 118 x 68 feet. Passengers: 776 Cabin class, 784 Tourist class, 579 3rd class.
Sixteen Steam Turbines manufactured by the shipbuilder, single reduction geared to quadruple screws. Speed 29 knots.

Launched 26.9.1934 by John Brown and Co. Ltd., Clydebank, for the company's North Atlantic services. Her keel had been laid on 27.12.1930 and the newbuilding became widely known by her Yard Number, which was No. 534. On 10.12.1931, work was suspended on account of the deepening depression and company losses. This caused national concern and it was eventually announced in 12.1933 that the British Government were prepared to advance funds for the completion of the vessel on the condition that Cunard and White Star Line amalgamated. This was effected in 2.1934, and construction commenced again 3.4.1934. In 8.1936 she crossed the Atlantic eastwards at an average speed of 30.14 knots and gained the Blue Riband from the French NORMANDIE. Although the French liner regained the record some months later, QUEEN MARY took both the eastward and westward records in 1938 and retained them until 7.1952. In 3.1940 she was taken up by the British Government for service as a troop transport and on 2.10.1942 she ran through the light cruiser CURACOA off the coast of Ireland. The cruiser sank within minutes with the loss of 338 of her crew. In 1946 she was returned to her owners and given a thorough overhaul at her builders. Sold in 1967 to the city of Long Beach, California, considerable sums were spent converting her to a maritime museum, hotel and convention centre. She arrived at Long Beach 9.12.1967 and is still there.

MAURETANIA (1938 - 1965) Steel Steamship.
35739 gross, 20170 net. 739 × 89 × 51 feet. Passengers: 486 Cabin class, 390 Tourist class, 502 3rd class.
Six Steam Turbines manufactured by the shipbuilder, single reduction geared to twin screws. Speed 23 knots.

Launched 28.7.1938 by Cammell Laird and Co. Ltd., Birkenhead, for the company's North Atlantic services. She was to become the largest liner to use the Port of London. In common with other liners in the fleet, she was taken up in 3.1940 by the British Government for trooping duties and served in this role until 1946. She was then sent back to her builders for reconditioning and commenced her first post-war commercial voyage on 26.4.1947. From 1962 she was employed almost exclusively on cruising and her hull was painted light green. In 1965 she was sold to shipbreakers at Inverkeithing.

QUEEN ELIZABETH (1938-1969) Steel Steamship.
83676 gross, 42011 net. 987 x 118 x 68 feet. Passengers: 823 1st class, 662 Cabin class, 798 Tourist class.
Sixteeen Steam Turbines manufactured by the shipbuilder, single reduction geared to quadruple screws. Speed 29 knots.

Launched 27.9.1938 by John Brown and Co. Ltd., Clydebank, for the company's North Atlantic services. When almost complete, it was decided to transfer her to New York because of the threat of German air attack. She therefore sailed 2.3.1940 from the Clyde without formal trials, arriving five days later at New York. In 11.1940 she was taken up by the British Government for service as a troop transport and was sent to Sydney, Australia, for conversion. Returned to her owners 2.1946, she commenced North Atlantic sailings in 10.1946 following reconditioning at Glasgow and Southampton. In 12.1965 an extensive refit commenced at Clydebank, at which time air-conditioning was fitted and extensive alterations were made to the passenger accommodation. The demand for passenger sailings across the North Atlantic was however rapidly diminishing, and she was considered too big for cruising. Sold in 1968 to The Elizabeth Corporation (a company in which Cunard held the controlling interest), she was taken to Port Everglades, Florida, and after conversion, opened as a convention centre and tourist attraction. She was renamed THE ELIZABETH. The venture was not a success and in 1970 she was sold to Seawise Foundations Ltd. (a subsidiary of the C.Y. Tung group) and was renamed SEAWISE UNIVERSITY. Taken to Hong Kong, work commenced converting her to a floating university. On 9.1.1972, when nearing the end of her refit, she caught fire and capsized in Hong Kong harbour. Removal of the wreck commenced in 1.1974 and was finally completed in 1979.

ASIA (1946-1963) Steel Steamship.
8723 gross, 5077 net. 488 x 64 x 32 feet.
Two Steam Turbines manufactured by Richardsons, Westgarth and Co. Ltd., Hartlepool, double reduction geared to a single screw. Speed 15½ knots.

Launched 12.9.1946 by Sir J. Laing and Sons Ltd., Sunderland, for the company's North Atlantic cargo services. Sold in 1963 to Taiwan, she was renamed SHIRLEY, and was broken up in that country in 1969.
Sisterships: ARABIA and ASSYRIA.

VASCONIA (1946-1951) Steel Steamship.
7058 gross, 4757 net, 431 x 56 x 35 feet.
Triple Expansion 3-Cylinder steam engine manufactured by Harland and Wolff Ltd., Glasgow. Speed 9½ knots.
Launched 11.4.1944 by Short Brothers Ltd., Sunderland, as EMPIRE PENDENNIS for the Ministry of War Transport, London. Purchased 1.4.1946 for the company's North Atlantic cargo service and renamed VASCONIA. Sold in 1951 to the Vestey group, she served as FRESNO STAR and later, MILLAIS. In 1960 she was sold to Hong Kong owners as GROSVENOR NAVIGATOR before her sale to Taiwan shipbreakers in 1966.
Sistership: VALACIA.

VANDALIA (1947-1954) Steel Steamship.
7281 gross, 4493 net, 423 x 57 x 34 feet.
Triple Expansion 3-Cylinder steam engine manufactured by Joshua Hendy Iron Works, Sunnyvale, California. Speed 10 knots.

Launched 1943 by California Shipbuilding Corporation, Los Angeles, as the Liberty ship GRANVILLE STUART for the United States War Shipping Administration. Before completion she was bare-boat chartered to the Ministry of War Transport, London, and renamed SAMARITAN. Her management was allocated to Cunard White Star Ltd. At the end of the war when much of the Government controlled tonnage was being disposed of, Cunard purchased her 17.4.1947 for their fleet. She was renamed VANDALIA and placed on the North Atlantic cargo services. Sold in 1954 to Liberian flag operators, she was renamed SIDERIS and was broken up in 1971 at Barcelona in Spain.
Sistership: VARDULIA.

MEDIA (1946-1961) Steel Steamship.
13345 gross, 7480 net. 518 x 70 x 33 feet. Passengers: 250 1st class.
Four Steam Turbines manufactured by the shipbuilder, double reduction geared to twin screws. Speed 17 knots.

Launched 12.12.1946 by John Brown and Co. Ltd., Clydebank, for the company's North Atlantic services. In 1961 she was sold to Cogedar Line of Italy and renamed FLAVIA. They had her extensively rebuilt and also lengthened, and then placed her on their service to Australia. In 1968 she was transferred to a cruising role and remained under the Italian flag until sold to Panamanian flag owners in 1982 as FLAVIAN. Laid up at Hong Kong following her purchase, she remained there, was renamed LAVIA in 1986, and was gutted by a fire which broke out 7.1.1989. At the time she was undergoing renovation, and was towed into shallow water where she heeled over and sank. She was damaged beyond repair and was sold to Taiwan shipbreakers.
Sistership: PARTHIA.

PARTHIA (1947-1961) Steel Steamship.
13362 gross, 7393 net. 518 x 70 x 41 feet. Passengers: 250 1st class.
Four Steam Turbines manufactured by the shipbuilder, double reduction geared to twin screws. Speed 17 knots.

Launched 25.2.1947 by Harland and Wolff Ltd., Belfast, for the company's North Atlantic services. In 11.1961 she was sold to the New Zealand Shipping Co. Ltd., London, for their service from London to Wellington. Renamed REMUERA, she was refitted at Linthouse and made her first voyage to New Zealand in 6.1962. These were the final days of the large passenger liners, and she only remained in their colours for two years. In 1964 she was sold to the associated Eastern and Australian S.S. Co. and was renamed ARAMAC. In 1969 she was sold to Taiwan shipbreakers.
Sistership: MEDIA.

CARONIA (1947-1968) Steel Steamship.
34183 gross, 18767 net. 687 x 91 x 48 feet. Passengers: 581 1st class, 351 Cabin class.
Six Steam Turbines manufactured by the shipbuilder, geared to twin screws. Speed 22 knots.

Launched 30.10.1947 by John Brown and Co. Ltd., Clydebank, for the company's North Atlantic services. From the outset she was employed cruising out of season, mainly from New York. In 1965 her passenger accommodation was modernised at Belfast, and a new open air swimming pool was fitted. Plans were considered in 1968 to sell her to Yugoslav interests for service as a floating hotel, but these were abandoned and she was sold to Panamanian owners as COLUMBIA. There then followed a refit at Piraeus after which she was renamed CARIBIA. She returned to cruising, but following an explosion in the engine room was laid up at New York on 25.3.1969. In 1974 she was sold to Taiwan shipbreakers but whilst in tow for Kaohsiung in 8.1974 went aground in Apra Harbour, West Guam and broke up.

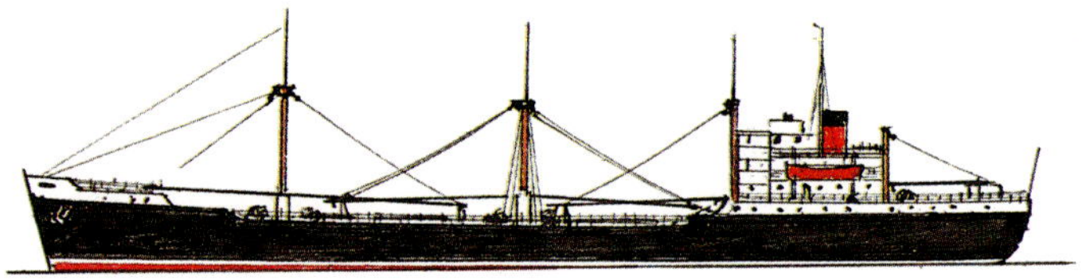

BRESCIA (1947-1966) Steel Motorship.
3817 gross, 2165 net. 326 x 50 x 26 feet.
6-Cylinder oil engine manufactured by Nordberg Manufacturing Co., Milwaukee, Wisconsin. Speed 11 knots.

Launched 1945 by Consolidated Steel Corporation Ltd., Wilmington, California, as HICKORY ISLE for the United States War Shipping Administration. She too was bare-boat chartered to the Ministry of War Transport, London and on 29.4.1947 was purchased by the company for their Mediterranean cargo services. Renamed BRESCIA, she remained in the fleet until 1966 when sold to Panamanian owners as TIMBER ONE. In 1970 she was sold again and converted for service as an oceanographic research ship and mining vessel. Renamed DEEPSEA MINER, she was employed recovering minerals from the seabed until sold to Spanish shipbreakers in 1974.

ASSYRIA (1950-1963) Steel Steamship.
8683 gross, 5013 net. 488 x 64 x 31 feet.
Two Steam Turbines manufactured by Richardsons, Westgarth and Co. Ltd., Hartlepool, double reduction geared to a single screw. Speed 16 knots.

Launched 19.1.1950 by Swan, Hunter and Wigham Richardson Ltd., Wallsend, for the company's North Atlantic cargo services. Sold in 1963 to Liberian flag owners, she was renamed LAERTIS, and HOLY TRINITY in 1970. Under her final name of DROMON, she was sold to Taiwan shipbreakers and was broken up in 1974 at Kaohsiung.
Sisterships: ARABIA and ASIA.

PAVIA (1953-1965) Steel Motorship.
3411 gross, 1828 net. 384 x 49 x 23 feet.
4-Cylinder Doxford oil engine manufactured by D. Rowan and Co. Ltd., Glasgow. Speed 13 knots.

Launched 19.3.1953 by William Hamilton and Co. Ltd., Port Glasgow, for the company's Mediterranean cargo service. In 1965 she was sold to Greek owners as TOULA N, and in 1974 passed to Taiwan shipbreakers.
Sisterships: PHRYGIA and LYCIA.

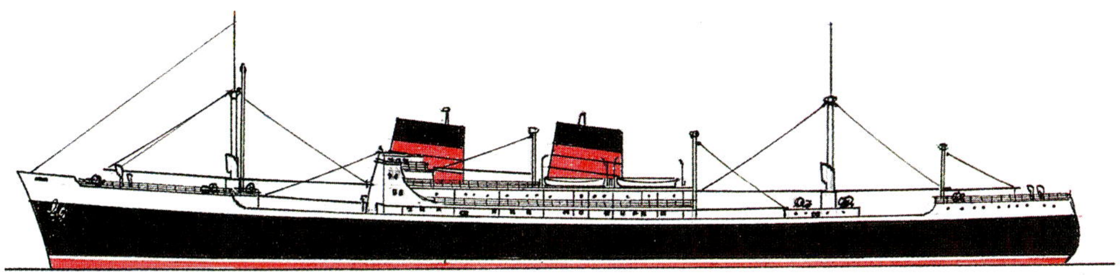

ALSATIA (1951-1963) Steel Steamship.
7226 gross, 3882 net. 483 x 64 x 28 feet.
Three Steam Turbines manufactured by Parsons Marine Steam Turbine Co. Ltd., Wallsend, double reduction geared to a single screw. Speed 16 knots.

Launched 30.11.1947 by J. L. Thompson and Sons Ltd., Sunderland, as SILVERPLANE for the round-the-world service operated by Silver Line Ltd. of London. Purchased in 1951 for the company's North Atlantic cargo services, she was renamed ALSATIA. In 1963 she was sold to China Union Lines, a subsidiary of the C. Y. Tung group, and was renamed UNION FREEDOM. Sold in 1977 to shipbreakers at Kaohsiung. The forward funnel was a dummy, and housed the ship's radar, monkey island, chartroom, radio-room, pilot's room, wheelhouse and captain's quarters.
Sistership: ANDRIA.

SAXONIA

SAXONIA/CARMANIA (1954-1973) Steel Steamship.
21637 gross, 11303 net. 608 x 80 x 28 feet. Passengers: 110 1st class, 833 Tourist class.
Four Steam Turbines manufactured by the shipbuilder, double reduction geared to twin screws. Speed 21 knots.

Launched 17.2.1954 by John Brown and Co. (Clydebank) Ltd., Clydebank, for the company's Canada service. On 27.9.1962 she returned to her builder for an extensive refit, during which time her hull and superstructure were painted in shades of green. Renamed CARMANIA, she was returned to the Canada service, but was employed cruising during the winter months. In 1967 she was transferred to permanent cruising and her hull was painted white. Withdrawn from service in 10.1971, she was laid up pending sale. She was eventually sold in 8.1973 through Panamanian interests to the U.S.S.R. and was renamed LEONID SOBINOV. She is still in service.
Sistership: IVERNIA, later FRANCONIA, CARINTHIA and SYLVANIA.

CARMANIA

ANDANIA (1959-1969) Steel Steamship.
7004 gross, 3501 net. 456 x 63 x 39 feet.
Two Steam Turbines manufactured by D. Rowan and Co. Ltd., Glasgow, double reduction geared to a single screw. Speed 15 knots.

Launched 19.12.1959 by William Hamilton and Co. Ltd., Port Glasgow, for the company's North Atlantic cargo services. She was transferred in 1969 to the associated Brocklebank Line, renamed MACHARDA, and retained on the North Atlantic. Sold in 1971 to Panamanian flag interests, she was renamed HUMI MAHIS. She passed to companies associated with the People's Republic of China in 1973 and was renamed YUNGJIAN, and later HONG QI No. 107. She is presumed to have been scrapped some years ago.
Sistership: ALAUNIA.

ALAUNIA (1960-1969) Steel Steamship.
7004 gross, 3501 net. 456 x 63 x 39 feet.
Two Steam Turbines manufactured by D. Rowan and Co. Ltd., Glasgow, double reduction geared to a single screw. Speed 15 knots.

Launched 12.5.1960 by William Hamilton and Co. Ltd., Port Glasgow, for the company's North Atlantic cargo services. She was transferred in 1969 to the associated Brocklebank Line, renamed MALANCHA, and retained on the North Atlantic. Sold in 1971 to Panamanian flag interests, she was renamed YUNGMING, and later HONG QI No. 108. She is presumed to have been scrapped some years ago. (Her funnel was different to that of ANDANIA to ensure better smoke and soot clearance).
Sistership: ANDANIA.

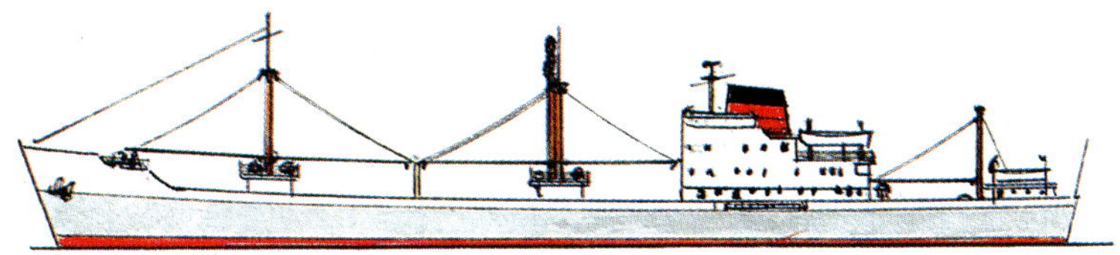

NORDIA (1961-1974) Steel Motorship.
4560 gross, 2432 net. 377 x 50 x 24 feet.
6-Cylinder Gotaverken oil engine manufactured by the shipbuilder. Speed 15 knots.

Launched 13.4.1961 by A/B Lindholmens Varv., Gothenburg, Sweden, as NORDIC for Rederi A/B Nordic, Sweden. Chartered by Cunard before completion for the North Atlantic cargo services, she sailed on her maiden voyage from Liverpool to New York as NORDIA. In 1974 the charter finished and she was sold to Cyprus flag owners as TEVEGA. Turkish controlled, Panamanian flag owners purchased her in 1981, when she was renamed DIMMER. She finally passed to Pakistan shipbreakers in 1987 and was broken up at Gadani Beach.

MEDIA (1963-1971) Steel Motorship.
5586 gross, 2658 net. 400 x 60 x 37 feet.
7-Cylinder Sulzer oil engine manufactured by Hawthorn, Leslie (Engineers) Ltd., Newcastle. Speed 16½ knots.

Launched 20.6.1963 by John Readhead and Sons Ltd., South Shields, for the company's North Atlantic cargo services. With the rapid advance of containerisation, she was found to be unsuitable for the trade and was sold in 1971 to the Western Australian Coastal Shipping Commission of Fremantle. Employed on the Australian coastal trade she was renamed BEROONA. In 1978 she became PALM TRADER under the Greek flag, and on 18.10.1983 sustained heavy fire damage to her bridge and engine room when lying at Bandar Abbas. Repairs were considered too expensive and she was declared a constructive total loss.
Sisterships: PARTHIA, SAXONIA and IVERNIA.

SAXONIA (1963-1970) Steel Motorship.
5586 gross, 2658 net. 400 x 60 x 37 feet.
7-Cylinder Sulzer oil engine manufactured by Hawthorn, Leslie (Engineers) Ltd., Newcastle. Speed 16½ knots.

Launched 17.10.1963 by John Readhead and Sons Ltd., South Shields, for the company's North Atlantic cargo services. In 9.1970 she was sent to Swan Hunter Shiprepairers Ltd. at North Shields and lengthened by 80 feet. Upon her return to service she was transferred to the services of the associated Brocklebank Line and was renamed MAHRONDA. Briefly chartered to Chr. Haaland of Norway in 1972, she was renamed CONCORDIA FOSS, but was soon renamed MAHRONDA again. In 1977 she was sold to Singapore interests as NEW DEER, finally passing to Chinese shipbreakers in 1983.
Sisterships: MEDIA, PARTHIA, and IVERNIA.

SCYTHIA (1964-1970) Steel Motorship.
5837 gross, 2829 net. 420 x 60 x 37 feet.
6-Cylinder oil engine manufactured by Sulzer Brothers Ltd., Winterthur, Switzerland. Speed 16 knots.
Launched 25.8.1964 by Cammell Laird and Co. (Shipbuilders and Engineers) Ltd., Birkenhead, for North Western Line (Mersey) Ltd., Liverpool. This company was a subsidiary of the shipbuilder, and chartered the vessel to Cunard on a long-term bareboat basis for their London-New York cargo service. In 1969 she briefly passed into Cunard ownership before her sale to T. & J. Harrison of Liverpool as their MERCHANT. In 1979 she was sold to Greek owners and was renamed SISAL TRADER. Whilst under arrest at Mayotte Island, N. of the Malagasy Republic, she ran aground 11.4.1984 during a cyclone. She was refloated and eventually towed to Mombasa where she was laid up until sold to Pakistan shipbreakers in 1986.
Sisterships: SAMARIA and SCOTIA.

QUEEN ELIZABETH 2 (1967-   ) Steel Steamship, later Diesel-Electric.
65863 gross, 38244 net. 887 x 105 x 56 feet.
Passengers: 564 1st class, 1441 Tourist class. When cruising 1400 one class.
Four Steam Turbines manufactured by John Brown Engineering (Clydebank) Ltd., Clydebank, double reduction geared to twin screws. Speed 28½ knots. Re-engined in 1986 with nine MAN-B&W oil engines driving electric propulsion motors.

Launched 20.9.1967 by Upper Clyde Shipbuilders Ltd., Clydebank, for the company. The contract had been signed on 30.12.1964 with John Brown and Co. (Clydebank) Ltd. (later to become a member of Upper Clyde Shipbuilders) and construction commenced 5.7.1965. The contract was valued at £25,427,000. Her entry into commercial service had been planned for 1.1969, but as a result of serious machinery problems on a shakedown cruise to the Canary Islands, Cunard refused to accept delivery and she was returned to her builders. Delivery was finally effected on 18.4.1969 and she sailed 2.5.1969 on her maiden voyage to New York. At that time, she had the traditional black hull, but a white funnel. In 5.1972 she was victim of a bomb scare when the company's New York office was informed that six bombs had been placed on board and a ransom of £140,000 was demanded. Four bomb-disposal experts were parachuted into the sea and taken on board, and the threat was found to be a hoax. Later in 1972 she was given an extensive refit which altered the passenger accommodation and many of the public rooms. On 1.4.1974 a fractured oil feed pipe resulted in a complete power failure during a Caribbean cruise. It was necessary to charter a Norwegian cruise ship to take off the 1642 passengers and land them in Bermuda. After extensive repairs in Bermuda she resumed her transatlantic schedule. Further engine problems were experienced when off the Scilly Isles 23.7.1976, resulting in serious damage to the starboard engine. On 4.5.1982 she was requisitioned by the British Government for service as a troopship during the Falklands Campaign. She sailed 12.5.1982 from Southampton with 5 Infantry Brigade, who disembarked 27.5.1982 at South Georgia. She then returned home, and after a refit resumed commercial service 8.1982 with a new shell-grey hull (later repainted black) and red, black ''topped'' funnel. During 1986 her steam turbines were removed and replaced by nine MAN-B&W oil engines driving two electric propulsion motors. The work was undertaken by Lloyd Werft, Bremerhaven, and the opportunity was taken to give her a complete overhaul with considerable internal improvements. She was redelivered to Cunard 25.4.1987. She is still in service.

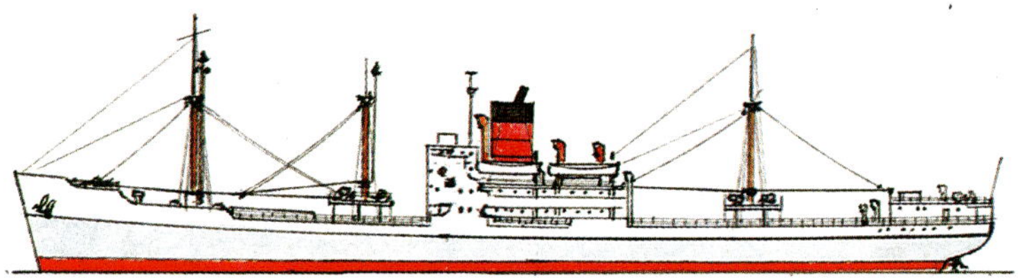

ASIA (1967) Steel Steamship.
3649 gross, 1519 net. 350 x 53 x 32 feet.
Triple Expansion 3-Cylinder steam engine with low pressure turbine with hydraulic coupling and double reduction gearing, manufactured by Central Marine Engine Works, West Hartlepool.

Launched 8.6.1956 by Caledon Shipbuilding and Engineering Co. Ltd., Dundee, as ALMERIAN for the Ellerman and Bucknall S.S. Co. Ltd. She was employed on the services of the associated Ellerman and Papayanni Lines of Liverpool. In 1962 she was renamed CITY OF LEEDS, and in 1964 reverted to ALMERIAN. First chartered by Cunard in 1966 for their Great Lakes service, she was renamed ASSYRIA. Within months she had reverted to ALMERIAN again, but in 1967 she was once again chartered by Cunard, this time as ASIA. Soon returned to Ellerman as ALMERIAN, she was sold to Cyprus flag interests in 1969 as THEOKLETOS. In 1974 she passed to Chinese shipbreakers.

ASSYRIA (1967-1968) Steel Motorship.
3422 gross, 1539 net. 345 x 52 x 32 feet.
8-Cylinder Burmeister & Wain oil engine manufactured by Harland and Wolff Ltd., Belfast. Speed 14 knots.

Launched 29.3.1949 by Harland and Wolff Ltd., Glasgow, as AMARNA for Moss Hutchison Line Ltd., Liverpool. Chartered by Cunard in 1967 for their Great Lakes service, she was renamed ASSYRIA. The charter ended in 1968 whereupon she was renamed AMARNA and returned to Moss Hutchison services. In 1975 she was sold to Katianav Shipping, Greece and was renamed KASTRIANI III. Sold again in 1982, the name of MONTROSE was suggested but never registered, and in 1984 she was delivered to Pakistan shipbreakers.

CUNARD CHAMPION in Cunard-Brocklebank colours

CUNARD CHAMPION (1973-1978) Steel Motorship.
15448 gross, 11221 net. 562 x 73 x 46 feet.
6-Cylinder Sulzer oil engine manufactured by Astilleros Espanoles S.A., Manises. Speed 15½ knots.

Launched 7.4.1973 by Astilleros Espanoles S.A., Bilbao, for worldwide bulk trading. In 1978 she was reported sold to the Onassis group who proposed renaming her OLYMPIC HARMONY. The sale was not completed and she passed to Philippine interests as EL CHAMPION. In 1984 she was sold to Sea Dove Shipping of Liberia, was renamed OCEAN PINE, and in 1985, FRIENDLY ISLANDS. She is still in service.
Sisterships: CUNARD CAMPAIGNER, CUNARD CARRONADE, CUNARD CARAVEL, CUNARD CALAMANDA, CUNARD CAVALIER, CUNARD CARRIER and CUNARD CHIEFTAIN.

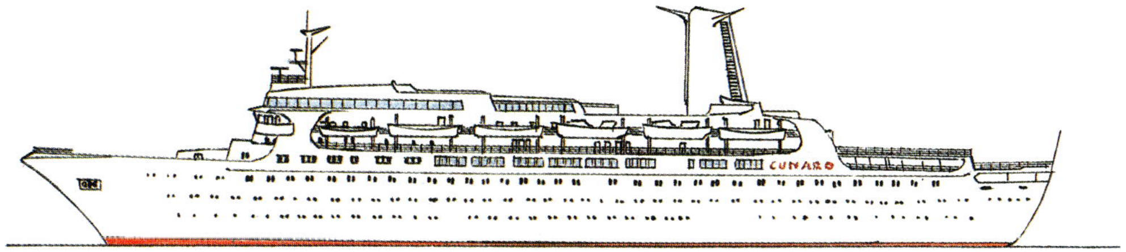

CUNARD ADVENTURER (1971-1977) Steel Motorship.
14151 gross, 8135 net. 413 x 70 x 41 feet. Passengers: 832 Cruise class.
Four 12-Cylinder oil engines manufactured by Stork-Werkspoor, Amsterdam, driving twin screws. Speed 21½ knots.

Launched 2.2.1971 by Rotterdamsche Droogdok Mij. B.V., Rotterdam, for Caribbean cruising. She had been ordered by Overseas National Airways in which Cunard held a considerable shareholding. In 7.1970 the company took full control of the contract. In 1977 she was sold to Klosters Rederi, Norway, and was renamed SUNWARD II. Following a refit at Bremerhaven she resumed Caribbean cruising. She is still in service.
Sistership: CUNARD AMBASSADOR.

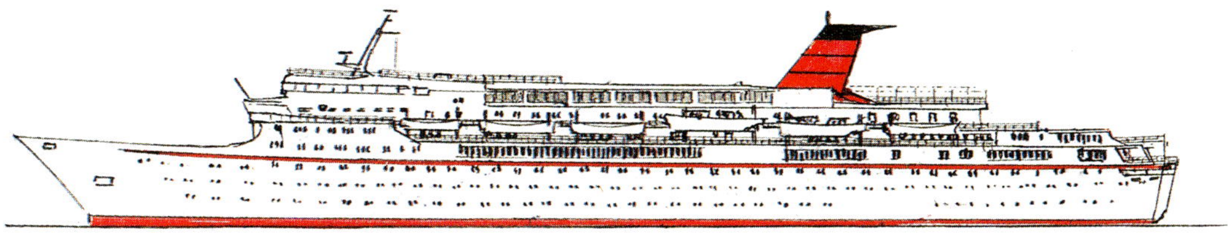

CUNARD COUNTESS (1974-) Steel Motorship.
17495 gross, 10096 net. 459 x 75 x 50 feet. Passengers: 950 Cruise class.
Four 7-Cylinder Burmeister & Wain oil engines, manufactured by Hitachi Zosen, Japan, driving twin screws. Speed 21½ knots.

Launched 20.9.1974 by A/S Burmeister & Wain's Skibsbyggeri, Copenhagen, for Caribbean cruising. Fitting out was contracted to Industrie Navali Meccaniche Affini, La Spezia, and was completed in 7.1976. In 10.1982 she was chartered to the British Government for service between Ascension Island and Port Stanley, Falkland Islands, and was returned in 5.1983 to Cunard. Following a refit at Malta she returned to commercial service. She is still in service.
Sistership: CUNARD PRINCESS.

ANDANIA (1976-1981) Steel Motorship.
6680 gross, 3638 net. 431 x 59 x 37 feet.
9-Cylinder Sulzer oil engine manufactured by G. Clark and N.E.M. Ltd., Wallsend. Speed 23 knots.

Launched 11.6.1972 by Smith's Dock Co. Ltd., South Bank, Middlesbrough, as GLASGOW CLIPPER for a subsidiary of Maritime Fruit Carriers of Israel. The majority of this fleet was purchased in 1976 by Cunard, who renamed this vessel ANDANIA. She was employed on worldwide reefer services. In 1981 she was sold to Greek interests and renamed EUROPA FREEZER. In 1986 she was sold to other Greek owners and was renamed BALMAR, and is still in service.
Sisterships: ANDRIA, ALSATIA and ALAUNIA.

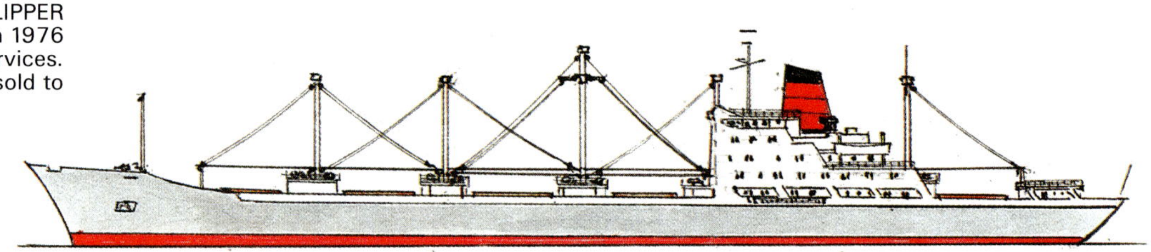

SAMARIA (1976-1986) Steel Motorship.
12059 gross, 6915 net. 534 x 75 x 44 feet.
9-Cylinder oil engine manufactured by A/S Burmeister & Wain's Maskin-og Skibsbyggeri, Copenhagen. Speed 23½ knots.

Launched 31.8.1972 by Aalborg Vaerft A/S, Aalborg, as CHRYSANTEMA, for a subsidiary of Maritime Fruit Carriers of Israel. The majority of this fleet was purchased in 1976 by Cunard, who renamed this vessel SAMARIA. She was employed on worldwide reefer services. In 1986 she was sold together with the bulk of the reefer fleet, to Kappa Maritime Ltd., London, and was renamed CAPRICORN. She is still in service.
Sisterships: SCYTHIA, SERVIA and SAXONIA.

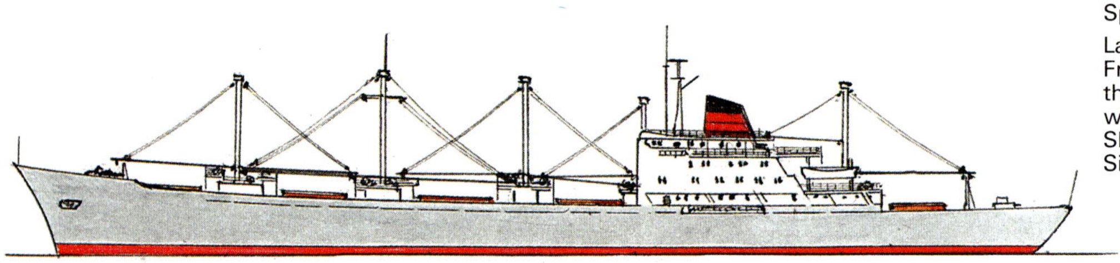

CARINTHIA (1976-1985) Steel Motorship.
9749 gross, 5703 net. 471 x 70 x 41 feet.
9-Cylinder Burmeister & Wain oil engine manufactured by the shipbuilder. Speed 20 knots.

Launched 19.1.1973 by Nylands Verksted, Oslo, as CANTALOUP, for a subsidiary of Maritime Fruit Carriers of Israel. The majority of this fleet was purchased in 1976 by Cunard, who renamed this vessel CARINTHIA and employed her on worldwide reefer services. In 1985 she was sold, together with the bulk of the reefer fleet, to Kappa Maritime Ltd., London, and was renamed PEGASUS. She is still in service.
Sistership: CARMANIA.

SAGAFJORD (1983-) Steel Motorship.
24108 gross, 13820 net. 551 x 80 x 55 feet. Passengers: 509 Cruise class.
Two 9-Cylinder Sulzer oil engines manufactured by the shipbuilder, driving twin screws. Speed 20 knots.

Launched 13.6.1964 by Forges et Chantiers de La Mediterranée, La Seyne, for Den Norske Amerikalinje A/S, Norway. In 10.1980 she was sent to Blohm & Voss, Hamburg, for a refit, and shortly afterwards Leif Hoegh & Co. acquired a substantial shareholding in the vessel. Purchased in 5.1983 by Cunard, her name was retained but the funnel has been repainted in Cunard colours. She is still in service.

VISTAFJORD (1983-) Steel Motorship.
24116 gross, 8907 net. 550 x 82 x 55 feet. Passengers: 670 Cruise class.
Two 9-Cylinder Sulzer oil engines manufactured by G. Clark and N.E.M. Ltd., Wallsend. Speed 20 knots.

Launched 15.5.1972 by Swan Hunter Shipbuilders Ltd., Wallsend, for Den Norske Amerikalinje A/S, Norway. In 1980 Leif Hoegh & Co. acquired a substantial shareholding in the vessel. Purchased in 5.1983 by Cunard, her name was retained but the funnel has been repainted in Cunard colours. She is still in service.

POSTSCRIPT

In the John H. Isherwood Collection there are drawings of several Cunard ships in subsequent ownership. Three of these are shown below.

ALBANIA, page 37, as she appeared in the 1930s as **CALIFORNIA,** owned by Navigazione Libera-Triestina S.A., Italy.

MEDIA, page 51, as the Cogedar Line **FLAVIA** in the 1960s.

PARTHIA, page 51, as the New Zealand Shipping Company's **REMUERA.**

INDEX

The names of ships illustrated in the book are shown in capital letters, e.g. QUEEN ELIZABETH 2.

In many cases, several ships were built to a similar design and the names of these ''sisterships'' to those illustrated are given in upper and lower case, e.g. Abyssinia.

An asterisk - * - against a ship's name indicates that Mr. Isherwood's detailed history of her was published in his series ''Steamers of the Past'' in the magazine ''Sea Breezes''.

The date after the ship's name is her year of build.

A further drawing of **MAURETANIA** of 1938 (see page 48)

The World Ship Society

This is one of over seventy books on shipping subjects which have been published by the World Ship Society. The Society was founded in 1947 and has grown to become the leading organisation for all who are interested in ships. It publishes the monthly journal ''Marine News'' which features the latest news on the world's merchant ships, warships and auxiliaries, together with articles on a wide range of subjects. The World Ship Photo Library is a major source of ship photographs, and the Society's Central Record is a unique and accessible source of information on ships dating back to 1830.

For details of the Society, and a sample copy of ''Marine News'' please send 30p in stamps (or four International Reply Coupons) to Department CP, 58 Bicton Street, Exmouth EX8 2RU, Devon, England.